T0240516

AutoUni – Schriftenreihe

Band 119

Reihe herausgegeben von / Edited by
Volkswagen Aktiengesellschaft
AutoUni

Die Volkswagen AutoUni bietet Wissenschaftlern und Promovierenden des Volkswagen Konzerns die Möglichkeit, ihre Forschungsergebnisse in Form von Monographien und Dissertationen im Rahmen der „AutoUni Schriftenreihe" kostenfrei zu veröffentlichen. Die AutoUni ist eine international tätige wissenschaftliche Einrichtung des Konzerns, die durch Forschung und Lehre aktuelles mobilitätsbezogenes Wissen auf Hochschulniveau erzeugt und vermittelt.

Die neun Institute der AutoUni decken das Fachwissen der unterschiedlichen Geschäftsbereiche ab, welches für den Erfolg des Volkswagen Konzerns unabdingbar ist. Im Fokus steht dabei die Schaffung und Verankerung von neuem Wissen und die Förderung des Wissensaustausches. Zusätzlich zu der fachlichen Weiterbildung und Vertiefung von Kompetenzen der Konzernangehörigen, fördert und unterstützt die AutoUni als Partner die Doktorandinnen und Doktoranden von Volkswagen auf ihrem Weg zu einer erfolgreichen Promotion durch vielfältige Angebote – die Veröffentlichung der Dissertationen ist eines davon. Über die Veröffentlichung in der AutoUni Schriftenreihe werden die Resultate nicht nur für alle Konzernangehörigen, sondern auch für die Öffentlichkeit zugänglich.

The Volkswagen AutoUni offers scientists and PhD students of the Volkswagen Group the opportunity to publish their scientific results as monographs or doctor's theses within the "AutoUni Schriftenreihe" free of cost. The AutoUni is an international scientific educational institution of the Volkswagen Group Academy, which produces and disseminates current mobility-related knowledge through its research and tailor-made further education courses. The AutoUni's nine institutes cover the expertise of the different business units, which is indispensable for the success of the Volkswagen Group. The focus lies on the creation, anchorage and transfer of knew knowledge.

In addition to the professional expert training and the development of specialized skills and knowledge of the Volkswagen Group members, the AutoUni supports and accompanies the PhD students on their way to successful graduation through a variety of offerings. The publication of the doctor's theses is one of such offers. The publication within the AutoUni Schriftenreihe makes the results accessible to all Volkswagen Group members as well as to the public.

Reihe herausgegeben von / Edited by
Volkswagen Aktiengesellschaft
AutoUni
Brieffach 1231
D-38436 Wolfsburg
http://www.autouni.de

More information about this series at http://www.springer.com/series/15136

Steffen Heinrich

Planning Universal On-Road Driving Strategies for Automated Vehicles

 Springer

Steffen Heinrich
Wolfsburg, Germany

Dissertation, Freie Universität Berlin, Department of Mathematics and Computer Science, 2017

Any results, opinions and conclusions expressed in the AutoUni – Schriftenreihe are solely those of the author(s).

AutoUni – Schriftenreihe
ISBN 978-3-658-21953-6   ISBN 978-3-658-21954-3   (eBook)
https://doi.org/10.1007/978-3-658-21954-3

Library of Congress Control Number: 2018940659

Printed on acid-free paper

This Springer imprint is published by the registered company Springer Fachmedien Wiesbaden GmbH part of Springer Nature
The registered company address is: Abraham-Lincoln-Str. 46, 65189 Wiesbaden, Germany

Acknowledgements

First and foremost I would like to thank my adviser Prof. Dr. Raúl Rojas. I want to express my gratitude for giving me the opportunity, tools and trust to pursue research in the field of robotics as a young undergraduate student. His support and attitude to continuously challenge the state of the art has help me to successfully finish this thesis.

I was very lucky to spend my time as a PhD student at two unique places for automated vehicle research. My sincere thanks also goes to my supervisor and adviser at Volkswagen Group Research, Dr. Arne Bartels and Dr. Simon Grossjohann, for their guidance, support and thoughtful advice on automated vehicles. I thank my mentor Dirk Langer and supervisor Jörg Schlinkheider at Volkswagen Electronic Research Lab in Belmont, California, as well as my colleagues Jake Askeland, Somudro Gupta and Jason Hardy for a great time of pioneering autonomous driving technology.

I specifically want to thank the research teams at AutoNOMOS Labs in Berlin and Volkswagen Automotive Innovation Lab (VAIL) in Stanford for their support and their valuable input as experts. In particular, I am grateful to Prof. Dr. Daniel Göhring, Daniel Seifert and Tinosch Ganjineh for being my early advisers and great team leads.

I thank Prof. Dr. Raúl Rojas, Dr. Arne Bartels, Jannes Stubbemann and André Zoufahl for being co-authors of my publications.

I want to thank my fellow PhD students Robert, Bennet, Tobias, Daniel, David and Patrick for our countless round tables and reviews. I thank my students André, Sebastian, Alex, Jorit, Christian, Jannes and Tobias for their excellent research. I am grateful for all feedback provided by Patrick, Jens, Peter, Fritz, Tobias, Jorit and Christian during the final stages of the thesis.

Most important to the pursuit of my PhD study has been the constant support of my family. I would like to thank my parents for their encouragement and unconditional support as well as letting me tinker with technology early on. And most importantly, I thank my wife and daughter for their patience, love and optimism during this challenging time.

Steffen Heinrich

Table of contents

List of Figures

List of Tables

Summary

This thesis describes a motion planning system for automated vehicles. The planning method is universally applicable in on-road scenarios and does not depend on a high-level maneuver selection automaton for driving strategy guidance. The majority of additional driving modes will therefore be processed in one particular processing unit for future high and full automation systems.

The thesis pursues three research questions that have given rise to all the solutions presented. What are the capabilities of a universal method? How can perception quality benefit from motion planning? And how can robustness against motion uncertainty be ensured?

A planning framework using graphics processing units (GPUs) for task parallelization is presented. A method is introduced that solely uses a small set of rules and heuristics to generate driving strategies. It was possible to show that GPUs serve as an excellent enabler for real-time applications of trajectory planning methods. For further assessment, planning benchmark criteria have been derived and applied.

Like humans, computer-controlled vehicles have to be fully aware of their surroundings. Therefore, a contribution that maximize scene knowledge through smart vehicle positioning is evaluated. It can be shown through experimental results that areas marked as relevant are more often covered by sensors by minor adjustments of the initial optimal planning solution.

A post-processing method for stochastic trajectory validation supports the search for longer-term trajectories which take ego-motion uncertainty into account. This achievement is a gain in the trajectory's temporal validity.

Zusammenfassung

In dieser Arbeit wird ein System zur Bewegungsplanung automatisierter Fahrzeuge beschrieben. Die Planungsmethode ist universell in definierten Straßenszenarien einsetzbar und benötigt keine höhere Entscheidungsinstanz auf Manöverebene zur Vorsteuerung einer Fahrstrategie. Für zukünftige hoch- und vollautomatische Systeme wird die Mehrheit der Fahrmodi in einer zentralen Recheneinheit generiert.

Die Thesis folgt im Aufbau drei Forschungsfragen. Welche Möglichkeiten ergeben sich durch einen universellen Planungsansatz? Wie kann die Umfeldwahrnehmung eines automatisierten Fahrzeugs von der Bewegungsplanung profitieren? Und abschließend, wie kann eine längere zeitliche Gültigkeit einer Trajektorie gegenüber Eigenbewegungsunsicherheit sichergestellt werden?

Im Rahmen der Arbeit wird ein Planungs-Framework vorgestellt, dass zur parallelen Aufgabenbearbeitung auf *Graphics Processing Units (GPUs)* zurückgreift. Es wird eine Methode präsentiert, die mit wenigen Regeln und Heuristiken auskommt, um Fahrstrategien zu generieren. Es kann gezeigt werden, dass GPUs die Planung einer großen Anzahl Trajektorien ermöglicht und die Echtzeitanforderungen an die Implementierung erfüllt. Zur Bewertung der Planungsmethoden, werden Bewertungskriterien abgeleitet und bei der Konzeption des Planungs-Frameworks einbezogen.

Automatisierte Fahrzeuge müssen, wie menschliche Fahrer, über ein vollständiges Situationsbewusstsein in Verkehrssituationen verfügen. Hierzu wird eine Lösung evaluiert, die die Positionierung im Fahrstreifen anhand der Informationsgüte der Wahrnehmungsmodule bewertet. Die Testergebnisse zeigen, dass Gebiete mit relevanten Informationen durch geringe Manipulation der optimalen Planungslösung wesentlich großflächiger eingesehen werden können.

Ein nachgeschaltetes Verfahren zur stochastischen Bewertung, im Bezug auf Bewegungsunsicherheit des automatischen Fahrzeugs, unterstützt die Suche nach Trajektorien mit längerer Gültigkeit. Dieser Beitrag verbessert die zeitliche Validität von Trajektorien.

1 Introduction, motivation and structure of the thesis

The vision of self-driving vehicles has reached the mainstream as it promises nothing less than the next mobility revolution. The potential benefits of automating the most popular transportation vehicle include reduction of road fatalities and CO2 emissions, an improvement in road utilization and free time for the person who would previously have been driving the vehicle in high automation systems. Automated driving is being discussed by society, governments, regulators, customers and potential mobility service consumers. In addition to traditional car makers and automotive suppliers, the development is also being led by IT companies.

1.1 Motivation of the thesis

The passenger car is one of the most popular transportation vehicles in the world. In the modal split of inland passenger transport in the European Union, cars accounted for 83% of kilometers traveled in 2014 [1]. At the same time, 1.25 million road traffic fatalities were reported by the World Health Organization in 2013 [2]. Specifically, road traffic injuries were still the number one cause of death among people aged 15-29. Hence, the European Union renewed its commitment to reduce the total number of road fatalities by 50% within 10 years back in 2010 [3]. Advanced vehicle automation has the ability to prevent a majority of these traffic accidents and decrease the number of fatalities on the road significantly.

The degree of vehicle automation is addressed by a classification guideline by the SAE [4]. In the case of highly automated vehicles, SAE level 4+ systems offer access to mobility services for anyone anytime. This includes people who do not currently hold a driver's license. Hence, new user groups such as children, elderly and people with disabilities would have the chance to move independently around their community.

In addition, the time spent in the vehicle could be used for something entirely unrelated to driving. For example, Americans spend 700 hours per year in their cars, whereas Europeans drive 300 hours on average. These numbers emphasize the potential of conditionally automated vehicles (level 3 systems), that will allow limited side tasks for drivers. Vehicle owners thus gain a large portion of the time they spend in traffic. As an entry scenario for such technology, it is interesting to take a look at traffic jams. In 2015 German drivers spent 38 hours in traffic jams on average. In larger cities this increased to 73 hours (Stuttgart) [5]. By comparison, in the US the average time spent in traffic jams amounts to 42 hours [6]. Today, this time has to be spent actively behind the wheel driving or supervising ADAS systems.

© Springer Fachmedien Wiesbaden GmbH, part of Springer Nature 2018
S. Heinrich, *Planning Universal On-Road Driving Strategies for Automated Vehicles*, AutoUni – Schriftenreihe 119,
https://doi.org/10.1007/978-3-658-21954-3_1

In addition, fuel consumption and CO2 emissions increase with greater traffic congestion. Vehicle automation can potentially improve traffic flow and optimize road usage, and therefore reduce the amount of traffic congestion.

Vehicle automation has the potential to lead to a change in how vehicles are operated and to eliminate existing idle times. The relationship between the human and the machine thus changes. Vehicle automation will enable drivers to become users of self-driving vehicles as well as consumers of mobility services. Humans will take over the steering wheel solely for pleasure, whenever they like and traffic, weather and road permit. Yet, there is more than one unanswered question about this new technology.

Challenges of vehicle automation

The task of driving a vehicle consists of sequences of perception, interpretation and reaction that human drivers learn and master relatively quickly. Independent of the traffic situation, humans can adapt and combine their driving actions in order to solve new and thus unknown tasks. Therefore, decision-making on driving strategies needs a centralized approach, in contrast to the situation with many distributed specialized ADAS systems within one vehicle. Unfortunately, as are all other robotic tasks, automated driving is subject to the *Moravec paradoxon* [7]:

> It is comparatively easy to make computers exhibit adult level performance on intelligence tests or playing checkers, and difficult or impossible to give them the skills of a one-year-old when it comes to perception and mobility.

Current vehicles can already be equipped with a wide range of *advanced driver assistance systems (ADAS)*. Their task in respect of motion planning is to support a single driving task with high quality partial automation (e.g. traffic jam assist) as the driver is continuously supervising the system. For higher levels of automation it is necessary to cover more driving modes while further decreasing the total number of specialized systems. However, we can find specialized systems for higher automation levels (SAE level 3 and above [4]) as well. These ADAS systems have been built for one particular purpose, such as driving in traffic jams, construction sites or valet parking. Behavior adaption for unknown situations exceeds the system boundaries.

Level 3 systems require less driver interaction and Level 4+ systems must operate without humans as fallback options. This increases the requirements placed on self-driving subsystems such as perception, object fusion and tracking as well as mapping services. All are key to understanding context in an observed environment. The ability to perceive the world is influenced by sensor types, quality and setup positions, whereas traffic density and occlusion seriously affect perception.

As part of this thesis, a universal approach for generating driving strategies is implemented to address the challenges mentioned above. The planner optimizes trajectories for automated vehicles in on-road scenarios. It considers ego-motion uncertainty and optimizes for optimal sensor coverage.

1.2 Thesis research questions

An automated driving system will need to perform any driving task they have been designed for as well as an experienced human driver to live up to passengers' expectations. For this thesis, three research questions have been identified and their implementations addressed to contribute to the current state of the art. All research has been conducted at the department for *Driver Assistance and Integrated Safety* at Volkswagen Group Research Wolfsburg, Germany. Throughout the following chapters, these questions act as the central thread:

- Universal planning systems: What can a universal approach demonstrate? And what are comprehensive benchmark criteria?

  - A compound of specialized systems has likely redundant logic in the overall implementation.

  - Combining systems to form a new driving mode is impossible.

- Knowledge locality: How can perception benefit from motion planning?

  - External pieces of global knowledge (backend, V2X) are virtual sensors.

  - Optimizing driving strategies to maximize perceived information.

- Does modeling motion uncertainty support temporal validity of planning solutions?

  - Terrain, latencies or measurement errors - all lead to a discrepancy between prediction and real outcome.

1.3 Thesis contributions

This thesis presents a framework for concurrent trajectory optimization methods, later referred to as *Phase Space Planning (PSP)*. In contrast to existing methods, this works without any high-level automaton. The contributions to the motion planning state of the art can be summarized as:

Trajectory planning framework The software component integrates into the exist-ing automated driving architecture of *Volkswagen Group Research*. It is capable to generate universal on-road driving strategies concurrently on a modern GPU. It is mainly set up with a 7D state space (position, velocity and momentum), an at-tribute that was eponymous: Phase Space Planning (PSP)

Rule and heuristic based planning A single state planning system was an initial premise, in order to consolidate planning methods and to combine their abilities. All behaviors are derived from one objective function. Rules and heuristics are integrated into state space creation and sampling processes.

Smart positioning Similar to the philosophy of *active vision*, the vehicle is posi-tioned to maximize scene knowledge. Therefore, sensor coverage of relevant areas (manually selected) is maximized.

Handling ego motion uncertainty Trajectory post-processing for temporal robust-ness is presented. Long lasting trajectories are identified by using a *Linear-Quadratic-Gaussian method (LQG)* to estimate the likelihood of future collisions taking into account motion uncertainty.

Publications

Crucial parts of each contribution have been published in conference papers by the author of this thesis:

- Modeling complex behaviors for automated cars in phase space (2014, FISITA Automotive World Congress) [8]

- Real-Time trajectory optimization under motion uncertainty using a GPU (2015, IEEE IROS) [9]

- Framework zur Planung einer Fahrstrategie für automatische Fahrzeuge in Echt-zeit (2016, AAET) (in German) [10]

- Optimizing a driving strategy by its sensor coverage of relevant environment information (2016, IEEE Intelligent Vehicles Symposium) [11]

1.4 Thesis outline

The following chapters cover the motion planning state of the art and a general idea of the planning framework. As we take a closer look at implementations and experiments, the thesis follows the processing pipeline as shown in Fig. 1.1 from left to right through the planning architecture.

First, an introduction to motion planning is given in Chapter 2. Taxonomies and definitions of common concepts are described and the major aspects of planning

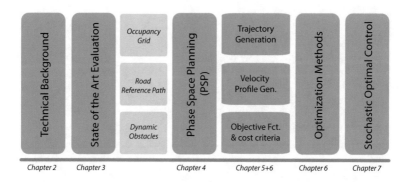

Figure 1.1: Schematic thesis outline following the data flow of the planning framework described in this thesis. The thesis gives background information and an overview about state of the art (left) before addressing the required fused sensor data representations (center, light gray). The core mechanisms of a sampling-based planning system (center, dark gray) and a post-processing stochastic optimal control method (right) are discussed and evaluated in detail.

in highly dynamic environments are highlighted. In Chapter 3 related work is discussed and the motion planning state of the art is presented. The overall idea and architecture of the planning framework project is described and discussed in chapter 4. Chapter 5 introduces technical implementations of the sampling-based planning system. It covers path and trajectory generation as well as the optimization process. Chapters 6 examines behavior creation by designing feasible components for the objective function. It presents the idea of a smart positioning behavior that optimizes sensor coverage of relevant environment areas. The fact that motion uncertainty always exists in a real world control problem is covered in Chapter 7. A stochastic optimal control approach is added to the planning pipeline in the form of a post-processing step. The thesis concludes with a summary and an outlook for future work and challenges remaining in the field of vehicle automation.

2 Preliminaries

This Chapter provides the theoretical background and is to be understood as a preface to the ideas contained in, and the contributions made by this thesis, which are presented later. First, an introduction to terms and nomenclature is given. Second, strict distinctions of automated driving as a special case of motion planning are emphasized. Third, a categorization of basic driving maneuvers for passenger cars is undertaken and definitions for each group are given. Finally, the general problem statement is presented.

2.1 Introduction to motion planning

The problem of *motion planning* is a major field of robotics as for any given task a robot has to *act* to accomplish its task. Motion planning is therefore a key component of almost every robotics domain such as intelligent transport, logistics or medical surgery. Motion planning facilitates automated manipulation tasks in human environments on land, in the air or under water. A very basic example of motion planning is the piano mover's problem. Here, it seems self-explanatory that in an environment (living room) filled with static obstacles (furniture) it can be challenging to move a large object with a complex geometry to its designated position. New generations of robots include not only complex compound bodies such as humanoids, but also grasping arms and animal-like robots. Even familiar machines such as ships, planes, drones, medical equipment or cars that have been around for decades have been given new levels of automation in their latest product releases.

Motion planning is sometimes understood as a form of collision avoidance. This is true for problem formulations that care solely about *if* an agent has arrived at its final pose, leaving out the aspect of *how* this was achieved. The answer to the first task is a classic solution to fixed boundary conditions. For the latter, a more vague formulation is needed to influence the overall shape of a solution - this is achieved by using so called *constraints*. Constraints are particularly important as soon as machines interact or share their environment with humans. In such a scenario, a basic collision free solution is not sufficient as this does not hold up to the expectations related to requirements such as driving comfort. Moreover, the quality and feasibility of the robotic motion is linked to these constraints (e.g. kinematic or dynamic models).

© Springer Fachmedien Wiesbaden GmbH, part of Springer Nature 2018
S. Heinrich, *Planning Universal On-Road Driving Strategies for Automated Vehicles*, AutoUni – Schriftenreihe 119,
https://doi.org/10.1007/978-3-658-21954-3_2

2.2 Terminology

For further reading a nomenclature is given in this Section to provide definitions of the most common conventions and terms that will be used throughout this work. Definitions that are relevant in a specific chapter will be introduced there.

Robot and space representations

In this thesis a robotic system $\mathcal{A}$ will be referred to as *vehicle* in the context of automated driving. Additionally, it is also called *agent* or simply *robot* in a more general setting. In all implementations, $\mathcal{A}$ will be reduced to a point mass (see Fig. 2.1), whereas the geometry of $\mathcal{A}$ is appended to the bounds of the obstacle landscape. An agent A acts in a Euclidean space $\mathcal{W}$, called workspace. This space is usually represented as $\mathbb{R}^N$ with N = 2 or 3. Within this space, position and shape of n obstacles $\mathcal{B}_i$, for $i = 1,\ldots,n$ are fully or partially known, depending on the given problem. An obstacle $\mathcal{B}_i$ is denoted by $\mathcal{B}^d$ when it is dynamic, e.g. when it is capable of moving. $\mathcal{B}_i$ is denoted by $\mathcal{B}^s$ for a static obstacle. A configuration space of a robot $\mathcal{A}$ is referred to as *C-Space* $C = \mathbb{R}^N$, for $N \in \mathbb{N}$. It is a manifold of usually the same number of dimensions as the robot's degrees of freedom. A mobile robot needs at least a single degree of freedom to deliberately change its state. The configuration space is a set of points describing valid configurations for the observed robotic system. A planning solution has to consist entirely of a series of points from this space and is derived by means of search. When considering system dynamics including velocity, acceleration as well as time, the term *configuration* does not apply and is changed to *state*. As a unique entity, a state $x = [x_0,x_1,\ldots]$ describes a pose, velocity and acceleration of an agent within a state space S at a fixed time. The connection between two states is called transition T_i. An example is shown in Figure 2.1, where $\mathcal{B}_i,\ldots, \mathcal{B}_q$ are fixed rigid obstacles.

Similar to a state space, there exists a space for agent actions as well. A *control space* is a set of points describing possible (e.g. feasible) inputs to the system. Given an initial state, applying a control input induces and defines a state transition.

Optimal control problem and planning strategies

The objective of an *Optimal Control* problem is to find a trajectory that is feasible for a robot and collision free given its geometry. A trajectory shall describe a transition from an initial state x_0 at time $t = 0$ to an identified final state x_f at time $t = t_f$ as shown in Figure 2.1. It is likely that there is not a single answer to this query and it is preferable to find the one that minimizes a chosen objective function $\mathcal{J}$. $\mathcal{J}$ can be considered as an overall rating, e.g. using energy, risk, time or comfort as its measurement unit.

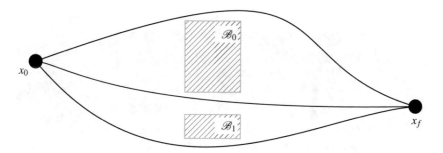

Figure 2.1: Instead of modeling the exact geometry of the robot, it can be simplified to a point mass representation x_0 in state space. As the actual shape is not neglected, but later added to all $\mathscr{B}_i$ objects in $\mathscr{W}$ in reduced detail.

In a more general form, the system can be described as a time discrete function

$$x = f(\dot{x}, u), \tag{2.1}$$

where f is a differential equation of the system kinematics. $\dot{x}$ is the predecessor state of x and u represents a control input to transfer $\dot{x}$ into x.

In contrast to other forms of autonomous robots, *autonomous vehicles (AVs)* explore their surroundings as they move at high speeds. It is unlikely that the whole area of application is entirely static, or even fully known a priori. This partial knowledge is due to perception blind spots, when obstacles occlude regions, as well as sensor resolution and range. It may furthermore change rapidly as dynamic objects move and interact as well. In respect of planning, the operating area is generally too large to cover at once. Moreover, it is computationally expensive to operate on the global route, while solving for a local problem. The same is true for infinite horizon problems where time $t_f \to \infty$. Hence, using a *receding horizon* approach, as shown in Fig. 2.2, adds a virtual horizon or way point between global target and current vehicle pose. This finite horizon is explored by an agent with a limited set of actions for a fixed amount of time t_f. During optimization, points close to the receding horizon have so-called cost-to-go shares that characterize the additional effort for a final transition to the actual destination. The exact costs might be unknown and a simplified heuristic is used.

Properties of motion planning results

The purpose of an optimization method is to gradually improve a solution to a given problem by adjusting its defining parameters. The quality is measured by an objective function as described in the preceding paragraph on *Optimal Control*.

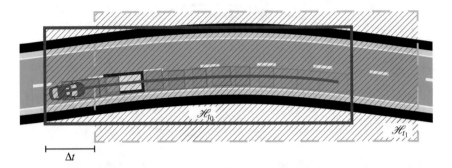

Figure 2.2: Splitting the planning tasks iteratively in smaller as well as local sub tasks $\mathcal{H}_{t_0}, \mathcal{H}_{t_1}, \dots, \mathcal{H}_{t_f}$ is called a receding horizon. The finite virtual horizon H_{t_f} that can be reached by the agent in a manageable amount of time, usually in seconds. It is explored by the agent with a limited set of actions for a fixed amount of time t_f.

Optimality is an important attribute of a motion planning algorithm. The *optimal solution* is the best of all possible variations with respect to the set objective. A solution is called *suboptimal* when the result obtained is very similar or close to the optimal solution. There might be more than one solution to the same optimization problem. Each solution is then called a local minimum in terms of the objective function. When a local minimum is found, it can be difficult for optimization algorithms to identify whether better solutions exist. A solution is a *global optimum* (or *optimal*), if and only if it is the best local minimum for this problem. Bellman defines a policy as a sequence of decisions, where a policy which is most advantageous according to some preassigned criterion will be called an optimal policy [12]. Bellman's general *principle of optimality* states:

> An optimal policy has the property that whatever the initial state and initial decisions are, the remaining decisions must constitute an optimal policy with regard to the state resulting from the first decisions.

In respect of the field of application discussed in this work the paradigm must hold for the result of the graph search.

In addition, trajectories must always instruct *feasible* motions that can be executed by the vehicle. However, a vehicle that is kinematically incapable of executing a required motion, runs the risk of colliding with obstacles or getting stuck in an insoluble position. This is shown in Fig. 2.3. In the worst case, this could lead to a robot system being trapped. *Feasibility* can be ensured in several ways and is linked directly to the entire control loop. Depending on the motion planning task the approximation of the vehicle dynamics model needs to be precise. This is not solely a trajectory generation problem. When building a search graph and connecting several trajectories to form a larger state transition, constraints must be met to ensure a smooth route across nodes. Therefore, the characterizing function has to be at least C^2 continuous. Using a precise approximation of the overall system is a desired method, but usually suffers from its complexity. In contrast to control sampling approaches, state sampling algorithms usually use spline or polynomial functions to approximate a feasible transition.

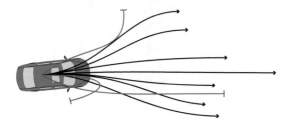

Figure 2.3: *Feasibility* of paths or trajectories with respect to the vehicle's kinematics refers to the capability of the automated vehicle to execute the given instruction considering all limitations and constraints.

It is up to the different manifestations of motion planning algorithms to ensure guarantees on their results and termination procedure. An algorithm is called *complete* if for an arbitrary initial condition a solution can be found in a finite amount of time, if such a solution exists. The demand for *completeness* is a common requirement for planning systems of autonomous vehicles. However, the introduction of heuristics to the algorithm addressing the problem statement can massively simplify tasks for a more efficient solution by weakening the performance guarantees [13]. A parameter of sampling based algorithms is the *density* of the points drawn. By picking more unique points, the sampling density increases and therefore the probability that the algorithms can report the positive answer (avoiding a false negative result) converges to one.

2.3 Taxonomy of planning methods

Over the past decades a large variety of planning methods have been presented for robotic applications. Finding the ideal approach for a motion planning problem is not straightforward as benchmarking depends on a great extent on the methods chosen. Small changes in the problem statement, the dynamic constraints, real-time or other non-functional requirements can have a big impact on the conceptional decision process. In this Section a brief overview is given about the most common methods and their respective fields of application.

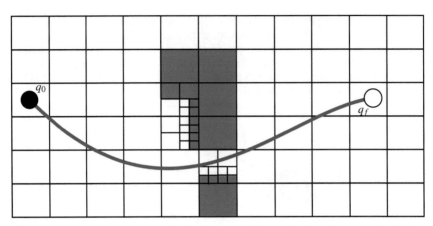

Figure 2.4: A region in state space can be discretized efficiently through a method called *cell decomposition*. It identifies regions of C_{free} and connects them to elements of the same occupation category. A collision check for a transition between configuration q_0 and q_f benefits from such representations.

Roadmap-based planning

In 1996 the idea of *probabilistic roadmaps* for path planning in high dimensional spaces was presented by Kavraki et al. [14]. The problem shown in Fig. 2.4 can be represented using a method called *cell decomposition* that identifies regions of C_{free} and connects them to neighboring elements of the same occupation category. A similar technique was initially presented by Nilsson [15] as a *grid model*. To build this representation of C_{free}, it has to be known explicitly. The results are corridors of C_{free} in the form of a graph representation. Hence, graph search can be applied to find paths through the space. An edge represents a collision-free sub-path through C_{free}. This structure contains a complexity reduced set of configuration sequences. Well known representatives of this planning category are: Retraction [16], exact and approximate cell decomposition [17] [18], visibility graphs (as described in [19]). In this thesis cell decomposition was implemented in the form of quadtrees, similar to the work of Frisken and Perry [20]. The authors focused on a memory-access-optimized and table-free methods.

Sampling-based planning

Motion planning under differential constraints is a computationally hard problem. The idea of only partially investigating the state space while still obtaining an optimal solution sounds too good to be true. Indeed, sampling-based planning does not give the same guarantees. The performance decreases in more complex search space settings (e.g. nodes might not be sampled well in narrow C-space passages). These methods are probabilistically complete, which means that the probability converges to 1 that a solution is found if and only if one exists. If no limit for the number of samples exists, the program may run forever. Sampling-based methods are widely used and well understood, though the problem for high dimensional C-spaces is still computationally hard. Another strength of sampling-based methods is their ability to execute fast queries as the state space is reduced to a small number of representative nodes. This concept is also applicable to complex C-space structures, which can be stored simply as a node collection. As a drawback it should be empathized that a low sampling density has its limits in state spaces with multiple obstacles.

Probabilistic methods

A way of avoiding the work of searching the whole configuration space is to approximate C_{free} by randomly drawing samples from C-Space and add them to a graph representation. Thus, completeness is traded for decrease in the run time. Algorithms of this category lack an explicit representation of C_{free} and new nodes are tested for collisions before adding the new transition to the configurations already explored. Two strategies have become very popular. First, *probabilistic road maps (PRMs)* [14] draw random samples and add them to their road map when a local planner rules out any collisions. The road map then links the new nodes to all neighbors that are accepted by the local planner. Second, *rapidly exploring random trees (RRTs)* initially have a similar strategy by randomly drawing nodes from state space until the tree reaches a target region. New nodes are added to an existing node in the tree structure by a local planner. In the event of a collision, a node close to the obstacle $\mathscr{B}_i$ is chosen as the new node in the tree. In every kth iteration, where k is preferably greater than 100, a node from the target region is chosen as the sample.

Artificial potential fields

A fairly different approach operates within an *artificial potential field* as a local planning strategy [21]. The C-space parts O_{static} and C_{free} are modeled as repulsive and attractive forces. The goal state is the biggest attractive force. Iteratively, a gradient descent is applied. Hence, the robot moves step-by-step in the direction of the sum of all forces that affect it at this position. A drawback of potential fields is that they need to represent C_{free} explicitly. Furthermore, they do not scale well with high dimensional problems and cannot guarantee the optimal solution as the robot might get stuck in local minima.

2.4 Motion planning for automated cars

The early prototypes of automated vehicles that operate on public roads in mixed traffic - cars driven by humans and machines - were already being developed several decades ago. Ernst D. Dickmanns and his team from the *Bundeswehr University Munich (UniBw)* demonstrated automated driving on highways in 1980. In contrast to ground vehicles, aerial, nautical and space vehicle automation has already become a reality, at least for basic operating modes. Exceptions are then reviewed or executed manually by a human expert. The majority of existing systems operate in open space with only a small number of static obstacles and no narrow corridors. This is because in narrow passages, drift, disturbance or minor control errors can be hazardous. In addition, aerial, nautical or space vehicles are constantly overseen by a human expert with special training. This eliminates the possibility of system misuse, such as an operation outside the system boundaries or the infringement of an obligation to supervise the system.

The problem of a self-driving system is shown in Fig. 2.5. An agent has to deal with static and dynamic obstacles in a structured environment (lanes) while approaching a construction site with a speed limit. The system needs to perform an accurate localization and pose estimation as well as a perception and tracking task. For general routing a detailed digital map has to be present as well.

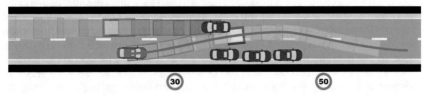

Figure 2.5: This Figure illustrates the problem statement of a self-driving vehicle. It has to deal with static and dynamic obstacles in a structured environment (road, lanes, traffic signs) while approaching a zone with parked vehicles protected by a speed limit. The system relies on an accurate localization and pose estimation as well as a powerful perception and tracking capabilities. In addition, a digital HD map is used for the general routing task on lane level.

A self-driving system operates in a highly dynamic environment that is challenging to predict. Other vehicles close to the agent move with high speeds and collisions must therefore be avoided at all times. The workspace of this problem can be simplified by adding knowledge about the general road structures to the system's world model. However, advanced driving modes have to be treated differently to the basic form of lane keeping and cruise control. This equally applies to parking and intersection tasks, which are not covered in this work.

2.4.1 Diversity of driving environments

Applying the rules of driving and adapting them to all sorts of still unfamiliar situations is an important skill of a human driver. This ability is hard to reproduce in automated driving systems. Every driving situation is shaped by the course of the road and therefore specifications for three main types of German roads are presented in this Section as an example. Each category differs in its road curvature, maximum speeds and lane width among others. These characteristics are derived from the German policies for road construction (*RAA* [22], *RAL* [23], *RASt* [24]) and have been extended by road type specific challenges for autonomous driving. The following paragraphs are not ordered according to complexity.

Highways The characteristics of highway driving in Germany are a minimal curvature, high vehicle speed and large differences in relative vehicle velocities. With lateral movements the offset changes slowly compared to the progress made in longitudinal dimension. These highways are designed with the largest vehicles (heavy trucks) in mind. The maximum size of a vehicle is 2.55m x 16.5m (18.75m for a truck-trailer combination). Therefore a highway lane allows lateral movement of between 0.7 and 1.2m within the lane markings.

Rural roads Rural roads are defined in construction classes EKL 1 to 4. Compared to highways these roads have a maximum speed limit and tolerate higher curvature radii along shorter arc length paths. The infrastructure is more minimalistic, as there are sections without crash barriers. Rural roads can be constructed in ways that have less impact on nature. The range of vision is, however, less than that on highways as straight stretches are shorter and the road profile less planar.

Urban roads Road construction has a long history in urban areas and many exceptions exist because of this. The *Jungfernstieg* in Hamburg is Germany's oldest asphalt road built in 1838. Historic streets do not follow today's standards, e.g. roads consist of sharp turns and narrow corridors. This road type is therefore considered the most difficult environment in which to operate as drivers or automated systems must interact with all types of road users. There is a high probability of unforeseeable situations that require the ability to make allowance for traffic law infringements, e.g. pedestrians crossing the road when the light is red.

2.4.2 Planning assessment criteria and driving modes

The variety of existing planning systems for automated vehicles is large, and too divers and specialized for straightforward comparison. This thesis therefore develops a set of 10 assessment criteria, which make it possible to categorize existing methods and understand their effectiveness and drawbacks for the purpose of creating a universal driving strategy generator. The set was created with this special case for automated driving in mind. It may apply to other other disciplines, but should not be considered as a general method for motion planning assessment.

Assessment criteria

With planning systems, we differentiate between a global route and a emphlocal planning horizon (local sub-segment of the route). Storing global information wastes scarce resources (memory and run time). Neither the state of the environment nor the spatial occupation around the goal area is constant, and should therefore be ignored for short-term decision-making on local problems. Moreover, the minimal length of the planning horizon has implications for the resulting trajectory quality.

For *sampling strategies* one could naively pick states from C-space hoping that the problem could be solved efficiently. This approach wastes memory and ignores the fact that even the simplest knowledge about the system and its field of application can drastically decrease the need for a dense sampling coverage in all regions of the C-space. The sampling can be done randomly by picking candidate from C or C_{free}, with some pre-processing (e.g. a reference path, see below). With respect to the environmental structure prior knowledge from digital maps can steer the general distribution and local density of sampling points.

The sampling can be done in either the agent's *state* or *control space* (see Section 5.5.1) A system's state can be transferred into a new state as long as the process complies with constraints of the dynamics. The *state transition function* has to guarantee trajectory feasibility. State transitions can be organized in multiple layers and form longer, more flexible transitions when stitched together. The states of each layer can be seen as waypoints.

With respect to *optimality*, the state space structure as well as the sampling density (endless sampling vs. threshold) influence the optimality property of an approach. The parameters chosen can therefore directly affect run time. Planning systems for automated driving have constraints on *real-time execution* and the planner's run time. As soon as the traveling velocities increase and the scenery can spontaneously change and affect the robot's time-to-collision (TTC) the run time has to be known and guarantees have to be adhered to. A planning system has to ensure a minimal update frequency without any changes in quality at all times. Time exceedance is classified as system failure.

Planning can be performed in the whole state space or in prioritized areas. Using pre-processed *reference paths* as a guiding element for the sampling process sets the focus on close-structured states and known parts of the environment. Information can come from digital maps or lane detection algorithms. It is the used to guide the search in state space. The downside is, that the performance (run time and quality) is also dependent on the quality of a reference path and the uncertainty coming from sensor measurement noise. Similarly, the presence of *distortion* has an impact on motion control a real world application. The question for a motion planning system is whether it is integrated into the motion model hypothesis (and to what extent) or whether it is neglected. For some cases the latter approach works well, especially when the resulting errors are small enough to be compensated in the control loop.

Planning and control methods may be decoupled. The *integration of a control method* affects the overall performance. Planning for automated vehicles is a cyclic single query task. The result of such a query should ideally not differ significantly from the previous answer. The trajectory shall be valid for longer period of time.

Planning *tactical maneuvers* is an ability to provide motions that serve a higher-level purpose. For two agents to cooperate, for example, a beneficial value is optimized and ideally almost all those involved improve their traffic situation. The whole traffic situation should definitely benefit from this task [25]. The ability to adapt spontaneously to actions of other agents or humans by copying their behavior or partially integrating their plans into one's own driving strategy is a powerful method to integrate an agent into mixed traffic scenarios.

Driving mode classification

As new *Advanced Driver Assistance Systems (ADAS)* enter the market and new research products are published, the variety of problems addressed makes it hard to differentiate between the systems. Several institutions have made proposals on setting a framework and establishing a common language for automated system classification. Two major publications have been released by the German *Bundesanstalt für Straßenwesen (BASt)* and the *Society of Automotive Engineers (SAE)*. Both categorize systems according to their general degree of automation, operating modes and system supervision requirements. The SAE draft shown in Fig. **??** *"Taxonomy and Definitions for Terms Related to On-Road Motor Vehicle Automated Driving Systems"* [4] allows statement to be made about future systems. The SAE publication categorizes automated driving systems into 6 different levels: (0) manual driving, (1) driver assistance, (2) partial automation, (3) conditional automation, (4) high automation, (5) full automation.

Whereas levels 1 and 2 are to be monitored solely by the driver, who is therefore fully liable for the vehicle's behavior in traffic, levels 3 to 5 require a computer system to monitor the environment at all times. Level 0 describes a conventional manually driven system.

However, in respect of motion planning the *driving modes* category allows three values (none, some, all). The question of a complete set of driving modes for an automated vehicle is not yet defined. For this work it was therefore necessary to define clusters of *elementary driving* modes. These clusters shall cover all basic motion of an automated vehicle. It is then necessary to evaluate whether a universal planning system is capable of finding instructions to describe these motions.

Table 2.1: Levels of vehicle automation

Level	Title	Execution of steering/ acceleration	Environment monitoring	Fallback option	System capability
0	No automation	Human	Human	Human	n/a
1	Driver Assistance	Human	Human	Human	Some
2	Partial Automation	System	Human	Human	Some
3	Conditional Automation	System	System	Human	Some
4	High Automation	System	System	System	Some
5	Full Automation	System	System	System	All

Summary of the SAE's latest version on the levels for automated vehicles. This thesis reports on systems operated on level 3 and above. Table data based on SAE J3016: Society of Automotive Engineers, 2014 [4]

In the publications of Nagel et al.'s [26] [27] driving maneuvers have been identified and classified as either primitive or a complex sequence of primitive motions. On the basis of these categories the catalog was then refined by Toelle [28] with explicit modes for advanced driver assistant systems in mind. Generalization is used to group pairs of similar entities. Furthermore, the criterion for the clustering of the driving modes is free space coverage:

- Cluster 1: start, follow, approach
- Cluster 2: intersections
- Cluster 3: double lane change
- Cluster 4: lane change, parking (parallel)
- Cluster 5: take turns, parking (perpendicular)
- Cluster 6: U-turn

This thesis uses this clustering and the elementary maneuver definitions for differentiation. The free space maneuver coverage of each cluster is a good indication for the search space complexity.

2.5 Problem statement

The objective of the motion planning approach presented here is to find a collision free, smooth and feasible state transition for a universal driving strategy. The strategy is provided by a sampling-based planner that has a single state of operation (*drive*). It continuously tries to find a sequence of state transitions with minimal costs to reach a target on the planning horizon. A valid transition is defined as a sequence of states that connects an initial state ξ_I and a target state ξ_F within a goal region $\mathscr{S}_G \subset \mathbb{R}^m$ in a finite amount of time and satisfies all boundary conditions. It becomes a *motion planning problem under differential constraints* which belongs to the family of general *two-point boundary value problems (BVP)* [29]. The constraints are derived from obstacles present, limitations of vehicle dynamics and prior knowledge about the road topology. Therefore, trajectories are generated that are guaranteed to be smooth as well as feasible for the vehicle. As for all BVPs constraints are set for various intentions. The key task is to eliminate the parts of the state space that hold vehicle states that are infeasible, hazardous or represent areas that are classified as *not drivable* by a perception system or map.

Additionally, the uncertainty of the ego vehicle motion can be modeled with a disturbance variable ε_k. For Chapters 4 to 6 we assume an accurately measurable and undistorted motion of the ego vehicle. The system must indicate failures and identify exceptions as they appear and forward these to a safety unit.

3 Related work

This chapter gives an overview of the current state of the art in motion planning. It summarizes key aspects of publications that are relevant to the field. Additionally, it presents literature from other robotic domains. In the hierarchical model of *sense, plan* and *act* planning systems for automated vehicles operate in the latter two phases. Donges [30] transfers this model into the context of driving tasks and maneuvering and established three categories: *navigation, motion planning* and *stabilization.* This thesis focuses on motion planning and touches the aspect of stabilization in Chapter 7. Related work in the field of trajectory planning is covered in Section 3.1. Optimization methods are compared in Section 3.2 and Section 3.3 outlines the differences in driving mode selection. The classic task of navigation, in the form of routing to a destination, is not covered.

This chapter follows the general structure of the thesis. Technologies and algorithms are introduced and discussed in this order.

3.1 From advanced driver assistance systems towards automated cars

Work on automated cars has been published around the globe in the decades since the 1970s. However, the circle of developers and researchers has grown recently. The latest milestones of vehicle automation have been presented by universities, car manufacturers as well as technology. The driving modes and maturity levels presented differ, as does the behavioral complexity of the showcases. The solutions presented range from proof of concept to close to series production vehicles.

Today, the most advanced level 2 driver assistance systems can be found in the premium car segment. However, the mid-range product segment already offers solutions such as collision warning and avoidance, assisted parking, adaptive cruise control as well as lane keeping functionality at higher speeds. The implementations of this thesis are created in the context of other research and development projects of the Volkswagen Group, which presented three forms of motion planning tasks recently. In January 2015 Audi has demonstrated a long distance drive from San Francisco to the CES venue in Las Vegas. Audi's *piloted driving* vehicle "Jack" drove 550 miles (over 880 km) on American highways through two states. In contrast to other showcases, journalists were allowed to sit in the driver's seat and oversee the ride after a special training session. A few months earlier, self-driving racing cars of the VW Group were shown on the Hockenheimring. In 2016, the V-Charge project [31] demonstrated automated valet parking and charging solution. Car makers and companies such as BMW [32], Daimler [33], Volvo [34] and NVidia [35] have showcased similar projects lately among others.

© Springer Fachmedien Wiesbaden GmbH, part of Springer Nature 2018
S. Heinrich, *Planning Universal On-Road Driving Strategies for Automated Vehicles*, AutoUni – Schriftenreihe 119,
https://doi.org/10.1007/978-3-658-21954-3_3

Research related to this thesis has covered the geometrical representation of agents, optimization algorithms and data structures in equal measure. An overview of the current state of the art is presented in the following sections. The on-road driving task is covered in detail in this chapter.

3.1.1 Trajectory planning

Connecting two vehicle states in state space with an optimal and feasible state transition is a core functionality of a planning system. This problem is not limited to automated vehicles but is relevant to every robotic system. Many algorithms that are mentioned in this section have their origins in other domains.

To demonstrate a more complex driving experience the path generation problem is generally fitted for certain use cases. The common clustering is done by environmental premises and leads to three groups:

- on-road driving (structured dynamic environment)
- free form navigation (unstructured dynamic environment)
- parking maneuvers (unstructured, mostly static environment)

It can be assumed that all clusters have at least a drivable surface or corridor. Corridors are thus wide enough to maneuver through if a solution exists. Consequently, there are many motions that are not feasible or optimal.

When representing unstructured workspaces including obstacles, Elfes' approach of occupancy grids [36] has been used and steadily reused in the context of automated driving. The author used potential functions in combination with an occupancy grid for mobile robot navigation. The grid is a two-dimensional random field that stores stochastic estimates of the occupancy of the cells in a spatial lattice. Hundelshausen et al. [37] implement a "tentaclelike planner which operates in unstructured environments by means of an occupancy grid. The grid is continuously updated by lidar sensor measurements. Other publications such as Mouhagir et al. [38] and Scholz et al. [39] are derived on the basis of this idea. "Tentacles̈are trajectories with different curvature values, reaching different endpoints in global space. The collision free trajectory with its endpoint closest to that region is chosen as the control input.

In contrast to intelligent systems that transport neither people nor cargo, automated vehicles need to ensure smooth transitions between certain driving maneuvers. It is therefore not sufficient to react to a single measured snapshot of the environment. Trajectory planning must consider a deliberative approach for automated driving modes to ensure consistent decisions.

Trajectory planning using a reference path

There are two common ways of planning trajectories with a sampling method. Either the input is sampled (*control sampling*) or the estimated pose is set by the sampling algorithm (*state sampling*). State sampling along a guiding representation is an effective method to reduce complexity of the search space. Using lane information as the reference is a common method in literature [40] [41] [42].

McNaughton et al. [42] present a lane-centered state lattice approach. The highly parallel planning system for non-holonomic vehicles is capable of demonstrating a basic set of driving maneuvers, including following a lead vehicle, changing lanes and swerving around obstacles. The driving behaviors are characterized by the cost terms of the system. A lattice is created on the basis of a sampling strategy. This guarantees feasible state transitions throughout the graph. It has been used at the CMU as an internal benchmark. Publications of Gu et al. [43] [44] shift the focus onto adaptation and stability. The approach is twofold, as are previous methods [42] [45]. Temporal and spatial components of the trajectories are generated separately. The authors discuss challenges regarding trajectory continuity in earlier projects such as [42] and identify the reasons. Path planning is done rapidly by designing the reference curve as an elastic band.

Another high dimensional approach is introduced by Ziegler and Stiller in [46] and [47]. Their spatio-temporal lattice consists of seven planning dimensions covering position (2D), velocities (2D) and acceleration (2D) as well as a time component. Trajectories are generated as quintic splines, which have the attribute of being second-order continuous. This approach is similar to that of Pivtoraiko et al. [48] [49] and the work of Rufli and Siegwart [40]. These approaches share the state lattice structure of feasible motions and use kinematic vehicle models. The model subsequently transferred into a (2, n) chained form. Howard uses a high dimensional approach in combination with pre-computed control sets for rovers performing free form navigation in complex terrain [50].

A sampling-based inner-city driving application is presented by Schwesinger et al. [51], who use a reference path in a local vehicle-centered coordinate system. Thy furthermore use a suitable system model for numerical forward propagation and represent obstacles in a position and orientation relative to the vehicle. This concept was later used and refined by Li et al. [52], who use a Support Vector Machine to refine the reference path by maximizing the lateral offset to corridor boundaries. They use a *rapidly exploring random tree (RRT)* for optimization and clothoids for path descriptions. The inner-city scenarios in Zurich are simulated and exhibit a good performance even for discontinuous reference paths.

Non-graph-based on-road planning methods

The most common form of planner creates a large set of states from an initial vehicle pose for thousands of feasible target states. In contrast to the systems described
in the aforementioned publications, such planners do not rely on a graph representation as they identify the minimal costs as they create the samples. Alternatively,
planning algorithms create one solution and try to minimize their cost by deform-
ing the trajectory iteratively. Ruf et al. [53] present a framework that uses such an
approach. It is called *situation prediction and reaction control (SPARC)*. The system
is separated into two parts as the name suggests and applied to low speed driving
tasks (less than $50km/h$). The trajectory is developed using a Lagrangian method for
minimal damage along a path described by waypoints as its initial solution. *SPARC*
is a holistic approach, which considers all relevant goals as well as the dynamic limits of the ego vehicle. Their evaluation in simulations shows that the approach is
capable of handling a dynamic intersection scenario and static obstacles along the
lanes. However, it remains to be seen whether this concept can be applied to scenarios with higher speeds, more complex road topologies and an increased number
of traffic participants.

Werling et al. [41] [54] present a method for driving in well-structured environments. Their work covers motorways and inner city scenarios. A low-level and
a high-level instance operate in a road centric *Frenét frame*, where the former one
can be seen as the control mechanism and the latter encapsulates the trajectory optimization process. Lateral and longitudinal trajectories are therefore generated
separately as a 1D trajectory. The resulting set of swerve actions is described by
quartic and quintic polynomials and connects a goal state with the initial configuration. Werling et al. identify three different types of trajectories and sample
target states for them: Velocity, following and stop trajectories. The first describes
a maneuver without a target vehicle but with a limit for maximum speed. The second variant reacts to the driving behaviors of a target vehicle, whereas the third
describes a smooth state transition into a safe position where $v = 0$. Strategic sampling at the targeted goal area in a lateral and a time dimension allows the car to
react with fairly simple maneuvers. In a later work published jointly with Gutjahr,
Werling presents a similar approach [55], though one of which is well designed
for embedded systems. Gutjahr and Werling formulate their problem as a linear,
time variant, model predictive control problem. Similarly, Ziegler presents another
trajectory planning solution [56], which was used in the *Bertha* automated driving
showcase. For every perceived static obstacle a node is added and neighboring
nodes are then linked if their connection is not collision free. This identifies part
of the left and right edge of the road. Second, polygons are used to describe the
shape of parking cars or other static obstacles. A minimal distance and an orientation in parallel to the edge are then used as an initial estimate for the trajectory
optimization, thus transferring acceleration, acceleration rate, distance and angle
to the polygon set, an offset to the reference speed and change of direction from a
method of finite difference to an extreme value problem.

3.1.2 On-road swerve path generation

The task is to connect two vehicle poses (e.g. position and orientation) with a smooth, feasible and continuous transition. For a point-to-point boundary value problem (BVP), existing solvers provide polynomial and clothoid or spline shaped solutions. Moreover, trajectory shape, distance to obstacles and curvature rate of paths have a direct impact on driving comfort and maneuver acceptance. Dubins [57] presents an approach which is one of the early models for car-like agents. His proposition is that the shortest path between two points can be expressed by a combination of arcs with a maximum curvature and straight stretches. The model has to fulfill simple kinematic constraints assuming constant velocity. Reeds and Shepp extend this to the reverse motion in 1990 [58] and this is often applied to parking scenarios or trailer maneuvering. It can be used to express idealistic and jerky motions for an arbitrary goal pose. The problem of discontinuous transitions can be resolved by connecting arcs and straight lines with clothoid segments, as proposed by Fraichard and Scheuer [59][60]. A similar solution is proposed by Theodosis and Gerdes to derive race lines that are comparable to those followed by human professional racing drivers [61].

In 2003 Kelly and Nagy [62] presented a parametric optimal control solution for using clothoids to describe paths of non-holonomic vehicles in structured and unstructured environments. The clothoid is thus generalized such that the curvature κ is a cubic polynomial function of its arc length. This approach is later refined [42] with improved parametrization of the curvature function to avoid larger differences in magnitude of the parameter variables. This adds numerical stability to the gradient descent. In contrast, Levien [63] uses clothoids to describe fonts and characters. He discusses the efficient computation of clothoids and states that the problem "is considered solved as at least two implementations provide accurate results in a time comparable to that needed to evaluate ordinary trigonometric functions"[63]. In his thesis the author discusses different spiral forms and models. The constraints at a node are comparable to the ones that apply to planning waypoints. Polynomials of higher order (e.g. quintic polynomials) are necessary for approximations that consider the smooth change of steering at these nodes, for example. Given the curvature, orientation and position values can be derived. A basic clothoid is shown in Figure 3.1 where the function f is defined by $\kappa(s) = s$. The parameters for this approximation are then refined using a gradient descent.

As mentioned in the previous section Werling et al. [41] [54] use quintic polynomials for the trajectory formulation. In this approach the longitudinal and lateral planning is initially separated and later merged to obtain the final trajectory results. This allows a straightforward matrix multiplication procedure to be used to calculate a trajectory based on the current initial state and the target point. It is possible to use a path planner only, without considering time constraints. Wang et al. [64] propose a method based on an earlier DARPA challenge concept [65]. In this case Akima splines [66] are generated along a reference path. The time critical aspects

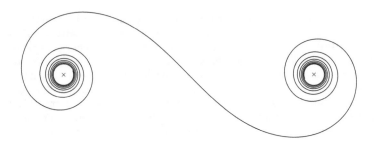

Figure 3.1: Illustration of an Euler spiral or clothoid

(e.g. collision avoidance and changing lanes) are handled separately on a higher planning layer.

In conclusion, the choice of swerve representation is directly coupled to the system requirements such as:

- state space dimensionality
- constraints on continuity (e.g. steering rate)
- feasibility and close dynamic model approximation

Generating velocity profiles

W. Xu et al. [45] propose a method of generating velocity profiles that refines a former lattice approach with constant acceleration [42]. However, speed and acceleration profiles are now retrieved by an inverse and acceleration profiles are continuous. First, the velocity space is discretized. Second, quartic instead of cubic polynomials are used to connect the initial state with its successors. This is to avoid discontinuities in the curvature rate of change at the beginning of a new planning cycle. The cubic polynomials are then used to connect all subsequent states. In this approach, Werling [41] adds a simple velocity profile to his proposed separation of lateral and longitudinal trajectory planning. The profile is added as soon as the two sets are combined.

3.1.3 Collision checking and avoidance

Avoiding collisions is a key requirement placed on motion planning algorithms. The state-of-the-art covers various ways of integrating and applying such a method in the context of automated vehicles. In general, collision checking is a computationally expensive task. Given the problem statement there are different ways to apply simplifications to the method to achieve better run time results. Paromtchik and Laugier [67] present an approach for a parking scenario that reduces obstacles to a few characteristic equations. Elfes uses potential functions in combination with an occupancy grid for obstacle avoidance [36]. In the case of sampling-based planners a common method is to incrementally draw state samples and generate transitions to these new states. Dolgov et al. [68] use such a method in the DARPA Urban Challenge for parking and vehicle maneuvering applications. The method of Ziegler and Stiller [69] for fast collision checking is designed for automated driving. In contrast to other motion planning use cases a vehicle geometry cannot be approximated by a simple disk model. A disk is rotationally invariant and does not describe the real rectangular shape of a car or truck very well. Ziegler and Stiller decompose the vehicle shape into a set of smaller disks. Thus, the overall geometry matches the actual vehicle with less spatial waste. The collision check can still be performed rapidly using the disk test several times for the set. In addition, massive parallelization can improve run time of a collision checking algorithm. Pan et al. [70] present an clustering strategy and a method to perform efficient collision queries simultaneously on GPUs.

Traffic predictions

Collision checks have to be similarly performed with static and dynamic obstacles. However, dynamic obstacles need to be predicted with respect to their current state and the environment.

Lawitzky et al. [71] predict a whole scene with all its participants including the mutual influence between traffic participants. Ruf et al. [53] create a large stochastic map based on the Kumaraswamy distribution. The behavior of dynamic objects is predicted into the near future and their presence is modeled as the potential degree of damage. The ego vehicle is therefore optimizing for an outcome with minimal damage. However, damage in this context does not refer to a crash but also to the likelihood of getting too close to an obstacle.

3.2 Optimization methods

Finding an optimal transition or sequence of transitions of vehicle poses in representations such as a set, grid, lattice, maze, graph or tree is address in well-known optimization algorithms. In the case of graphs, this can be achieved with general algorithms such as a breadth-first or depth-first search. Djikstra's algorithm [72] could also be taken into account, as dynamic programming. Graph search methods share parts of their methodologies, e.g. small modifications lead to new algorithm names, as in the case of D*. D* is presented by Stentz in [73] and [74]. D* is a modification of A* such that cost parameters can change during the exploration process. Dolgov et al. [68] [75] study practical search techniques in the context of the urban challenge and present a so called hybrid-A* for a navigation problem in unknown and unstructured environments. First, A* is used to find feasible trajectories. Secondly, a non-linear optimization is used to improve the initial solution. The proposition is that this frequently leads to the global optimum, but at least a local optimal solution.

This addresses in particular the exploration of state spaces that are unknown or partially unknown. A similar idea is used by Wahl et al. [76] by introducing heuristic model-based search space reductions for a dynamic programming method. Their planning approach focuses on predictive longitudinal control for hybrid electric vehicles. McNaughton and Urmson [77] introduce a focused A* heuristic recomputation (*FAHR*). It is considered an enhancement to an A* search whenever there are large differences between A* cost-to-go and the actual true cost function. FAHR supports the effort of escaping a local minimum by recomputing parts of the heuristic function while exploring the search space. Later, McNaughton et al. use an exhaustive search for optimization [42]. This is part of the *Tartan Racing team* project, the winning contestant in the DARPA Urban challenge. Others rely on similar methods. Stiller and Ziegler optimize trajectories for minimum squared jerk using an exhaustive search strategy [47] [46].

Random trees for search space optimization of automated vehicles are still a minority. As described for Schwesinger's contribution [51], RRTs are a powerful method that is able to operate in real-time scenarios. The RRT guides by random exploration of a high dimensional state space, which is another possibility to build and optimize for the best solution in a tree structure. In contrast to state space sampling, Schwesinger et al. use a single-track model for the car dynamics and are able to generate C^0 continuous paths. Earlier, LaValle and Kuffner [78] presented a technique to obtain control inputs from an RRT for hovercrafts in $\mathbb{R}^2$ and spacecrafts in $\mathbb{R}^3$ Tang et al. [79] saw the need for randomized planners to optimize high dimensional search spaces. They focus on the problem that randomized planners struggle in narrow, cluttered regions. They use local information to support the tree search in these environments. The authors Chen et al. [80] propose an exploration process guided by a heuristic search approach. The exploration method uses circles to walk through free space and stores all samples in a search tree. Moreover, the heuristic search part uses the circle-path as a heuristic to generate state transitions under kinodynamic constraints.

In comparison, Werling et al. also choose minimal jerk as their main criterion, thus ensuring smooth steering movements and acceleration profiles [41]. Later, together with Gutjahr [55] the optimization problem is then solved as a quadratic optimization problem with linear constraints of the vehicle's dynamic model. The combination of linear system modeling and a quadratic objective criterion provides an inexpensive way to add further constraints. This design allows the problem to be solved in state of the art QP solvers. Similarly, the overall optimization for [56] by

Ziegler et al. is performed by a sequential quadratic programming method instead of the previous exhaustive search approach.

Ruf et al. [53] connect two vehicle states and iteratively optimize the latest solution. Driving decisions are deformations to the trajectory and are made based on the information from a stochastic map (Kumaraswamy distribution). The map contains costs for the vehicle's presence for each position at a certain time (the temporal horizon is relatively short). The authors choose the cost unit of *damage*.

Task parallelization using GPUs

The workload of motion planning has several hot spots that can benefit greatly from massive parallelization on GPUs. First, large sets of sampling points need exactly the same algorithm for state transition generation. Second, collision checking for small time steps along a trajectory are computationally expensive. Third, evaluating trajectories by applying an objective function makes it necessary to process many similar operations for sequences of states. Fourth, optimization on huge graph structures can be partitioned into smaller problems. This enables huge sets of exploration results to be optimized with common methods such as A*, D*, dynamic programming or classical Djikstra algorithms in real time. Kider et al. [81] address this problem in their work and propose a randomized highly parallel implementation called R*. It guarantees probabilistic sub-optimality instead of A* deterministic optimality. Interestingly, this solution is highly parallelizable and is implemented and evaluated on GPUs. R* and R*GPU impose significantly smaller memory requirements compared to optimal methods. In contrast to other existing randomized planning solutions, Kider et al.try to find solutions with minimal costs and provide probabilistic guarantees on the quality whereas other planning systems focus on finding any solution to the given problem. For a classic variant of A*, Bleiweiss [82] finds an order of magnitude improvement using GPUs compared to a dual core CPU A* variant. The planning approach of McNaughton et al. [42] as well as the proposal of Hudecek et al. [83] use GPUs to optimize trajectories. GPUs have been used to solve and accelerate existing algorithms such as continuous-state Partially Observable Markov Decision Processes (POMDP). Lee et al. [84] benchmark a GPU variant against classic CPU implementation and find an improvement by a factor of 75 depending on the problem.

Besides generating trajectories, GPUs can also accelerate collision checking. Pan et al. [70] show how to check multiple configurations simultaneously for collisions. The algorithm performs a parallel hierarchy traversal for each collision query. The authors apply clustering techniques to bundle collision queries to cores. They introduce the notion of collision-packet traversal. This method ensures similar traversal patterns for all entities in the same bundle.

3.3 Driving mode selection

A *deliberative planning* approach considers a long-term target, the trajectory actually driven, as well as the current traffic situation. It forms a driving strategy for a planning horizon of several seconds that ensures a safe and comfortable cruise. In contrast, *reactive planning* approaches that rely solely on snapshots of the sensor data set without using any history or prediction can be very effective, e.g. the approach by Elfes [36]. Such a process is called *local planning*, where the sum of each planning step might lead into a dead-end or a weak global solution [38].

As stated in Chapter 2, there is more to motion planning than pure collision avoidance. In addition, tactical positioning adds value to the system and to all other basic driving maneuvers such as lane changing and lead vehicle following. Tactical motionhas several benefits:

- *hazard avoidance*, through cooperative driving behavior

- *perception gain* through active positioning

- *economy of time* through lane selection and number of lane changes

- *dynamic driving experience*, through acceleration for time gaps

The first two categories in particular are desirable for an automated vehicle that must consider driving defensively and with a high level of comfort. Cooperative maneuver planning is outlined and investigated by Düring et al. [25]. It is not covered in the planning framework presented here.

In the case of perception gain, the tactical objective is for the sensor coverage to include areas that provide valuable information about the scenery. The coverage, in contrast to the vehicle's general field of view (FOV), is the sector that is not blocked or hidden by obstacles.

This observation task is more common in other robotic domains, such as localization or mapping. It is known as *gaze control* or *active vision*, especially in relation to camera or lidar sensors. The idea is to take advantage of a directed, controlled robot gaze to reduce the uncertainty of sensor measurements for certain parts of the environment. This has proven to be useful for *Simultaneous Localization and Mapping (SLAM)* methods, such as that of Davidson and Murray [85] or Seara and Schmidt [86]. Saffiotti and LeBlanc [87] present an approach that creates a connection between representations of objects - in abstract and in the physical world of legged soccer playing robots. The technique of creating and maintaining this connection over time is called *anchoring*. Later, Kohlbrecher et al. [88] developed an algorithm to control the gaze of a humanoid robot for a camera-based environment perception. The gaze control helps to provide a more frequent update of the section of the obstacle occupancy grid that is more relevant for the robot's behavior planning. Similarly this is also under consideration for non-holonomic systems such as cars.

The actual sensor coverage is interesting for improving trajectory planning and for the vehicle's hardware setup. With sensor integration the sensor positioning has an effect on the perception outcome and gaze directions. Unterholzner and Wünsche show how it applies to the detection and tracking of road networks [89] [90]. In contrast, the approach adopted by Patels' et al. [91] for sensor adjustment of fast driving autonomous cars reduces the number of unobservable grid cells on the car's trajectory. Sensor data fusion can give a 360 degree field of view and a sensor measurement redundancy. However, this does not guarantee that the sensor coverage includes all relevant areas of information in complex structured environments, e.g. at intersections with pedestrian crosswalks. This idea is supported by the work of Plavšić's et al. on human drivers. Erroneous driving behavior is caused by lack of information due to focusing solely on the current lane. Obstacles in the peripheral view are perceived too late or not at all [92].

3.3.1 Multi-layered search space representations

Combining two or more state transitions in a multi-layered solution of a trajectory adds value to the degree of freedom with respect to expressing driving modes.

This is relevant as a strategy to handle longer time or special planning horizons. Adding waypoints or subnodes therefore reduces the respective length of a single transition. In the case of a quintic polynomial, this would be an additional swerve motion along the planning horizon. This can be found in [42], [43], [45] and [46]. Here, the authors use grids and lattices and connect transitions at representative nodes.

The vehicle motions are defined by a model predictive control solution solved in advance. In the work of Rufli and Siegward, the lattice is bound to the curvature of the road [40]. Hence, the authors point out that a globally fixed heading discretization as shown in [48] leads to a non-applicable lattice representation for structured environments. The proposed solution is a simultaneous generation of input and state lattice.

3.3.2 Hierarchical driving mode state machine

Instead of making decisions in the trajectory optimization process, it is common to implement a high-level behavior state automaton. This high-level planning instance sets parameters and activates local planning methods. Such a hierarchical approach is used in many implementations that have been discussed in this chapter. For example, Werling [41] uses a higher instance, called a *behavioral layer*, which handles long-term objectives such as lane changes. While this is beneficial for smooth and conservative motions, it prevents the system from choosing a more human-like maneuver, such as actively accelerating into a time gap for a lane change. Montemerlo et al. [65] introduce a trajectory planning system that relies on accurate localization methods and precise and detailed map data in order to successfully plan trajectories on a structured road network. Decision-making is not part of the planner. The system's driving behaviors are set in a finite state machine. A target region is chosen by a higher hierarchy module [36] [41] [38].

Ziegler's contribution [56] contains a hierarchical structure and driving behaviors are chosen top-down. Thus, the trajectory planning operates on the basis of these directions. First, the method identifies a corridor with the shortest route to the goal area at the end of the planning horizon.

3.3.3 End-to-end machine learning approaches for autonomous driving applications

Machine learning supported planning is nothing novel as such. In recent years and with deep learning [93] gaining more and more momentum in the field of pattern recognition, end-to-end driving solutions have been proposed for automated driving [35]. Enabling factors are the huge amount of data related to a problem that can be generated and stored as well as the advances in computing hardware. GPUs have become a driver of deep learning technologies.

Bojarski et al. [35] from NVidia propose an end-to-end framework using a Convolutional Neural Network (CNN) and three front facing camera images as inputs. The input is mapped on steering commands by learning a policy function with a neural network. However, learning policy functions has well known drawbacks. They are likely to suffer from unexpected behaviors due to the mismatch between a trained policy functions and the possibilities of states reachable by a reference policy. This problem has been addressed by Zhang et al. [94]. The authors propose a simulation based learning method that increases convergence and it is intended to rule out unexpected behavior. Santana and Hotz [95] propose a fully integrated solution, consisting of smart sensor (camera) and Recurrent Neural Network (RNN) capabilities in the backend, whereby the smart sensor acts also as the main generator of training data.

4 A framework for universal driving strategy planning

This chapter presents the overall planning philosophy of this thesis. In Section 4.1 the integration into the existing software ecosystem is outlined and it being a valuable source with respect to perception and control systems. The first research question of this thesis is about the benchmarking of planning systems. Exchangeability in defined stages of the planning process is therefore the main driver of the framework's architecture. The framework will be referred to as *Phase Space Planning (PSP)* throughout this thesis. All motion planning implementations created as part of this thesis, both CPU and GPU variants, comply with the PSP interfaces. The interface and representations are introduced in Section 4.2. By embedding PSP into the general software architecture, it seamlessly integrates with the latest research vehicle platform of Volkswagen Group Research. A newly developed visualization is part of the PSP package and is presented in Section 4.3. It serves as an evaluation and debugging module for fairly high-dimensional planning spaces.

4.1 Planning in high dimensional state space

Automated driving is a highly dynamic task. The vehicle and its environment have a certain momentum and inertia, which needs to be covered in the planning representation. A phase space is an extension to a C-space. It has more dimensions (velocity, acceleration and time) than the original configuration space [96] and thus covers the dynamic components and time dependencies in a system. A single point in phase space is therefore described by x and $\dot{x}$. Constraints in phase space often involve the time derivatives of acceleration or jerk, to express smooth transitions. It is widely accepted that increasing the dimensionality of the space is less computational effort than dealing with higher order derivatives [29]. In this thesis, the term state space will reffer to derived phase spaces.

The framework developed in this thesis is called *Phase Space Planning (PSP)*. It was designed to investigate and benchmark different sampling-based planning algorithms and to choose one solution for the final evaluations. The motivation was the fact that evaluation and comparison is a non-trivial challenge in motion planning. It is especially difficult when trying to compare two planning methods embedded in otherwise entirely unrelated research projects. The architectures vary in too many crucial specifications, preventing a fair comparison. For example, variations are of five kinds: (1) Variations in the amount and structure of environmental input data; (2) variations in the required model accuracy; (3) variations in the actuation latency; (4) variations in specific control strategy implementations; (5) variations in the hardware used.

© Springer Fachmedien Wiesbaden GmbH, part of Springer Nature 2018
S. Heinrich, *Planning Universal On-Road Driving Strategies for Automated Vehicles*, AutoUni – Schriftenreihe 119,
https://doi.org/10.1007/978-3-658-21954-3_4

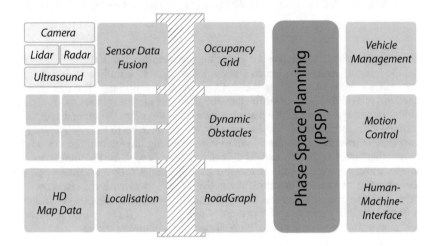

Figure 4.1: The PSP framework (dark gray) is embedded into an existing software ecosystem. It therefore supports all necessary interfaces to commonly used data representations. A digital map of the road, an occupancy grid map and a scene and object representation are PSP's link to the perceived world. It services the interface to the control unit with the latest trajectory planning result.

Essentially, a seamless integration into the existing software architecture, as shown in Fig. 4.1, was a key requirement for the design of the PSP framework. PSP therefore handles all existing incoming data streams and generates solutions in an established format which describes control targets. The sensor data is post-processed by perception modules. These modules fuse various sensor and vehicle data to provide an enriched format for the further processing modules on higher layers. Perception outputs include representations such as grid maps and detailed scene and object descriptions. The PSP framework collects data from these sources for each planning cycle, however, and calculates a prediction of these representations at the time the trajectory is expected to be executed. This structure and major ideas for this framework have been published in [8] and [9].

The inner core of the PSP framework was designed for the purpose of comparing and evaluating key components of the planning cycle. Modules such as *Path Generator*, *Velocity Generator* and *Optimizer* can be benchmarked against concurrent implementations as shown in Fig. 4.2, which also provides a detailed illustration of the system architecture. The state is described using 5 to 7 dimensions, which is a large space for vehicle maneuvering. A set of paths is generated along a reference line on the basis of the sensor data representations. This set is extended by a velocity profile generator. An optimization method is used to identify the combination of state transitions with minimal costs between initial and goal state. As a post-processing

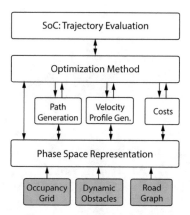

Figure 4.2: A diagram of the internal framework architecture: The data flow is bottom up. Based on the world representations, a set of paths is generated along a reference line. This set is extended by a velocity profile generator and evaluated by an objective function in the optimization phase.

step, the solutions are then used in a *stochastic optimal control (SOC)* process for a more stable result.

The following section will focus on specific aspects of the PSP framework and give detailed answers about the system design decisions.

4.1.1 Identifying key components of sampling based planning

PSP is a sampling-based planning framework. It is a common method, which LaValle described as the most successful motion planning paradigm in his 2004 book [29]. In its general form, the process of sampling-based planning follows three independent steps. This continuous planning cycle is illustrated in Fig. 4.3.

Initially, an *exploration* of the state space is performed. The number of samples and the clustering as well as the density of the drawing method are important parameters for the end result. Most samples are then connected through state transitions *generated* by another component. The final result is to be found as part of an *optimization* process minimizing an objective function.

The three components of the PSP framework are defined as follows. Each implementation for one of the planning steps must be in a generic form such that it is compatible with other methods running before or after that step. There are exceptions to this premise, e.g. efficiency reasons (memory management) for GPGPU implementations.

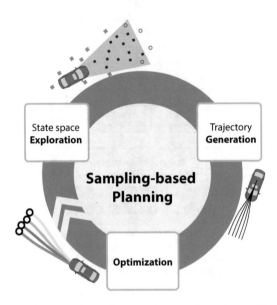

Figure 4.3: In its general form, a sampling-based approach can be partitioned into three key components. (1) Exploration phase; (2) Trajectory generation phase; (3) Search space optimization phase.

State space exploration A sample is drawn from the state space - either randomly or in a guided process. A guided sampling strategy follows fixed patterns or sets of rules and relies on expert knowledge. In the latter case, the quality of the sampling results has a strong dependency on the prior knowledge. Using expert knowledge allows the problem to be simplified, but there is a higher risk of not finding optimal solutions in the parts of the space that have been excluded by the drawing rules.

Trajectory generation All samples need to be connected to the current search graph (or the initial state at the beginning). These state transitions are highly constrained (e.g. vehicle dynamics). A trajectory model must be defined to approximate the real motion capabilities of the vehicle. A more exact model is more costly in terms of computing power and time. Each transition is independent of all the others in the same set.

Search space optimization We want to identify the best candidate (sample) with respect to the objective function, i.e. a vehicle state with the lowest transition costs in the set. Optimization methods suitable for this use case are well-known graph search algorithms, exhaustive searches (e.g. dynamic programming) or randomized algorithms (e.g. rapidly exploring random trees).

Not all sampling-based algorithms are suitable for the PSP processing pipeline. When sampling and transition generation is done as an atomic operation, it cannot be integrated into the framework. This drawback is still acceptable as most commonly known methods can be divided into these three components. And thus it is the enabler for a structure that allows the evaluation of single components and their variants. All methods presented in this thesis have been aligned with this structure.

4.1.2 PSP integration into existing architecture

The software architecture for automated vehicles at Volkswagen Group Research is built with a hierarchical planning structure in mind. The structure can be outlined in three layers:

- strategic (e.g. optimal lane-level routing),
- tactical (e.g. triggering lane change maneuvers),
- operational (trajectory generation and optimization).

The current planning systems therefore have a hierarchical state machine to identify the next tactical maneuver. On the operational level, several target points are sampled and subsequently compared. The system optimizes for the motion with the least squared jerk of each target point and chooses the one that involves the greatest braking force. The target points are set by separate tactical modules that handle a single behavioral task. Therefore, each behavior module uses perception data to identify regions well suited for its sampling purposes.

The perception modules derive higher-level representations of the vehicle's surroundings by fusing vehicle data and various sources of sensor data. The most important environment representation providers will be examined in detail. A scene $S = (G, O_s, O_d, \varphi)$ that describes the environment in a vehicle-centric way consists of a road topology G, an occupancy tri-state grid O_s (free, occupied, unknown), a sensor data fusion O_d tracking dynamic obstacles as well as the current vehicle pose $\varphi = $ (Position, Orientation). These environment representations are major products of the system's perception module and work as follows:

RoadGraph All edges representing lanes form a road network in a structured driving environment. The RoadGraph data structure presented by Homeier and Wolf [97] is shown in Fig. 4.4b. It describes all details of the topology, such as lane specific information: type of lane markings, speed limits, road curvature or adjacent lanes and driving directions. Data on traffic signs, traffic lights, lanes for cyclists as well as pedestrian crossings is stored for non-lane specific queries. But the RoadGraph is not a static data structure. It holds annotated up-to-date sensor

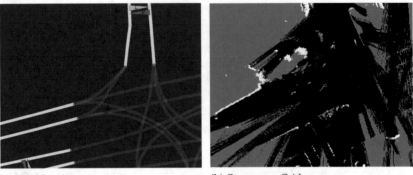

(a) RoadGraph   (b) Occupancy Grid

Figure 4.4: (Left) The road representation *RoadGraph* is a lane segment network with detailed information on attributes such as speed limits, traffic signs, road's curvature and more. (Right) Same perspective on the occupancy tri-state grid is
shown. Its aggregation of laser data is a main input for the planning process.
This grid stores all static objects in an occupancy grid with 10cm × 10cm grid
cells. (occupied = white, unknown = gray and free = black).

measurements, such as static or dynamic obstacles. This sensor-related information is automatically transferred into RoadGraph's coordinate system. The ego
vehicle is matched on this graph through localization.

Camera-based lane fusion So-called lane fusion is used for lane-relative localization and path prediction. The idea was published in Töpfer et al. [98]. Using
camera images and prior knowledge of the road structure, a spline is generated as
a path hypothesis. This reference is used by the planning system with additional
uncertainty information about the course of the curvature ahead.

Occupancy grid (static obstacles) The vehicle's lidar data is accumulated over time.
All static objects that have been identified in the sensor's field of view (FOV) are
stored in an occupancy grid data structure. The grid covers a region of 100m in
front and 25m behind the vehicle and provides a resolution of $10cm \times 10cm$. An
example of the occupancy grid is shown in Fig. 4.4a.

The existing planning system used a scene container that transferred this information into a lane centric *Frenet coordinate frame* [99] similar to [41]. In contrast to
this approach the PSP framework operates directly on the RoadGraph and grid
structures. Thus, the framework was integrated as a separate module between the
representations and the control instance, as shown in Fig. 4.5.

Both the RoadGraph and the grid are published at a frequency of 25Hz. The planning module operates at the same rate. PSP, in contrast, has a much lower planning

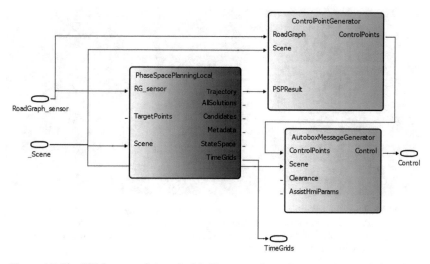

Figure 4.5: The PSP framework is embedded between the main perception modules. Their input is used directly to build the internal state space representation. It provides a trajectory and regularly transmits the result along the chain. *Control-PointGenerator* and *AutoBoxMessageGenrator* adapt input data and feed a control module in the vehicle. The RoadGraph provides upcoming road segments and is used as the reference line for sampling heuristics, for example.

frequency due to the complexity of the system. Therefore PSP always uses the latest data from these modules.

The control module receives the planning result from an internal *AutoBoxMessage-Generator* module. It has a general trajectory interface and operates at a frequency of 100Hz. The assumption is that the trajectory received is valid and contains enough information until the next package is delivered. At the beginning of each cycle a control point is extracted from the last known trajectory. This point shall be either 1.3s in the future or at least 12m in front of the vehicle. This point is used in separate units for longitudinal and lateral control. Apart from stabilizing the system, the control unit is also responsible for supervising the status of the vehicle interface and acts as a gateway for the human machine interface elements.

Target vehicle platform

The Volkswagen Group Research software architecture is independent of sensor types and vendors and was implemented independently of a vehicle platform. The

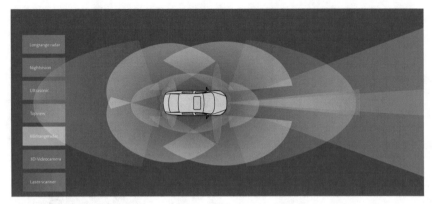

Figure 4.6: This illustration shows the sensor set and its coverage. Areas are colored
according to the related sensing device. Ultrasound sensors (red), radars
(white,blue), lidars (dark grey) and cameras (dark blue, orange) are the main
sensing devices of the test vehicle.

vehicle type and characteristics are set by parameters, as are the details on sensor type, features and setup position. A customized vehicle interface limits ranges
for compliant actuation forces comparable to current *Adaptive Cruise Control (ACC)*
and lane keeping systems [100]. The maximum deceleration is $3.5\frac{m}{s^2}$, when driving
above $60\ km/h$ and $4.5\frac{m}{s^2}$ when driving more slowly. The actuation force for steering
is limited to $3Nm$. See Table 4.1 for further details.

Table 4.1: Overview of the drive-by-wire interface limits

Force	Value	Unit	Notes
Acceleration F_a	2.5	m/s^2	
Deceleration F_d	3.5	m/s^2	$\geq 60km/h$
Deceleration $F_{d_{slow}}$	4.5	m/s^2	$< 60km/h$
Steering F_s	3	Nm	

The sensor set consists of lidar, radar as well as camera devices. The overall 360
degree sensor coverage is illustrated in Fig. 4.6 in more detail. The front is covered by a redundant setup of sensing devices of all types. All vehicles are research
prototypes and must be operated by trained drivers.

4.1.3 Challenges and opportunities: Modeling a universal drive

Current level 2 and 3 ADAS systems are decentralized implementations of driving
modes. Each system is either engaged by the driver by pressing a button or is part

of a subsystem that is active as soon as its master mode is active. ACC is a driver assistance (level 1) system that provides one driving mode of future motion planning systems. An ACC system itself consists of a hierarchical structure that uses sensor data to select which automaton state is most effective. Expanding the feature set is equivalent to adding new nodes to this state machine. Figure 4.7 sketches different approaches to designing a motion planning system. The generality increases from left to right. Whereas the maneuver is defined in detail in an ACC system, it is described as a trajectory for a common planning system. The trajectories are vehicle dynamic approximations and often simplifications. The state machine operates on maneuver tasks and parametrizes the specific trajectory generator. On the right, a system similar to PSP is presented. The driving task is unified for the very basic maneuvers. The state machine acts on a higher level for special motions such as safety critical swerves or complex intersection handling.

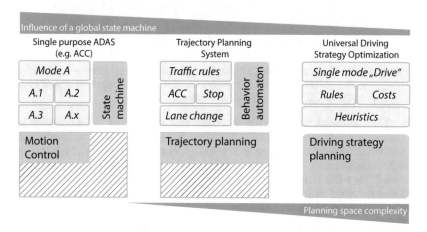

Figure 4.7: From left to right: Control and planning systems with an increasing amount of generality. An ACC system is of a hierarchical form, with a detailed description of internal driving modes. A state of the art trajectory optimization method uses dynamic models or approximations. It is controlled by a higher level planning automaton, which operates on maneuver tasks and sets parameters. On the right, a system similar to PSP is presented. The driving task is unified for the very basic maneuvers.

Trajectory planning and driving mode design merely follow a similar strategy. Figure 4.8 shows such a structure by way of example. Usually, there are three types of driving modes: (1) addressed driving modes, (2) unaddressed driving modes and (3) unknown and unaddressed modes. While *addressed* and *unaddressed* driving modes are system design decisions, unknown or unaddressed modes include all undefined system states or state transitions. In this case a system would operate outside its limits, in an unforeseen driving situation, similar to invalid trajectories that cannot be executed by a control unit.

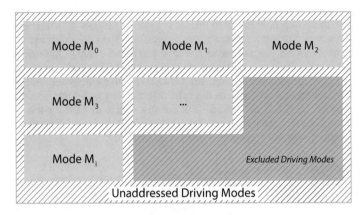

Figure 4.8: When designing a strictly hierarchical planning system, there are three categories of driving modes: (1) addressed driving modes, (2) unaddressed driving modes and (3) unknown and unaddressed modes.

The PSP framework is used to build monolithic approaches. The objective is to avoid explicit behavior implementation for two reasons. First, it is impossible to optimize for completeness. Second, explicit implementations have a high risk of redundant lines of code. When designing for a universal approach, the state machine-driven *If-Then-Else* structure is replaced by an objective function. Each component of this function is then responsible for a generic need such as lateral offset from the center of the road in the case of automated driving.

A single state-to-state transition described by a polynomial, for example, is limited in its swerve characteristics. Choosing a higher-order variant makes it difficult to control. A universal planning system for more complex driving modes therefore requires multiple temporal layers along the planning horizon. Each layer holds states that can be seen as waypoints towards the target area. The set of waypoints is stored in a search tree. Dynamic programming works well on such multi-level decision processes. It is a combinatorial search in a discrete search space, which guarantees a global optimal solution. It was therefore chosen as one of the optimization methods in the PSP framework.

4.1.4 Non-functional requirements: safety, comfort and acceptance

A motion planner for automated vehicles does more than collision checking and lane following. Its tasks are strategic with respect to driving experience (e.g. passenger acceptance), interactions with the traffic prediction and scene observation of other traffic participants. As for all planners in Volkswagen Group Research, the module must generate *safe* and *comfortable* trajectories. PSP uses a framework to validate methods that meet a wider range of non-functional requirements on automated driving such as *acceptance* and *inferability*. This supports the endeavors to answer the research questions about knowledge locality and *solution stability* raised in Section 1.2, which are both linked to a more human-like automated driving experience. *Human-like planning* is defined as a deliberative approach to imitate the adaptability of humans in complex traffic scenarios. Therefore, all PSP planners are comfort systems, where all maneuvers must operate within the boundaries of the circle of forces. All additional requirements are formulated in terms of the overall cost function within PSP.

Chapter 6 and 7 address non-functional requirements and contribute to the discussion about using motion planning to create benefits in earlier and subsequent modules (e.g. perception and motion control) along the process chain of a self-driving system.

4.2 PSP world representations

All motion planning systems need a form of virtual representation of the world. The PSP framework stores environment information such as a digital map and vehicle geometry in its state space. Future states of dynamic objects are predicted as they appear along the assumed path known from routing information. A state space consists of the following representations:

Vehicle state The vehicle geometry is hidden in its surrounding obstacle representations. Therefore, the vehicle geometry is added to the edges of static obstacle space. The ego- vehicle is thus described as a point mass. A state in phase space is a septuple (7-tuple) of position, orientation, velocity and momentum information.

Static object Static objects are derived from an occupancy grid. Front-facing lidar and camera systems contribute to this representation. Grid data close to road segments ($\leq 50m$) is extracted and stored in state space. At high speeds (e.g. on highways) PSP does not provide a detailed grid map.

Dynamic object Not all perceived dynamic objects have an impact on the final trajectory. In pre-processed filtering, objects are removed from the scene when their initial position is far from the road segment of the ego-vehicle or a neighboring segment. The remaining set of objects is predicted with respect to their velocities and orientations as well as the curvature of the matched road segment.

Reference line A reference line is created when the ego vehicle is matched on the road graph. The line is extracted from identified road segments as well as its successors. Its length is set on the basis of the current state of the vehicle and its options to act (e.g. max. acceleration). The reference line thus describes the planning horizon.

Vehicle data The vehicle status provides more than the basic vehicle state information such as velocity, acceleration and yaw angle. Indicator state and ACC set speed limit are also used in PSP, among others. The automated driving system is not released for all kind of roads and traffic types. Segments are therefore excluded and marked for manual driving by means of a geofencing method. This information (distance and time) is used by PSP to hand over the vehicle in sufficient time.

The world representation is reduced and simplified to the bare requirements of a motion planning system. Even though information is rejected, it is still important for other modules to keep it in the data set. As an example, object tracking might set up objects that are far away at the beginning, but within seconds they reach an area that forces the system to interact with them. So they might be excluded in the first planning cycles, but this changes as soon as they appear on a neighboring road segment.

Representation of space and vehicle localization

It is fairly expensive to process planning algorithms when the coordinate frame of the vehicle configuration space is not suitable for the application, especially when frequent coordinate transformations are necessary. In the case of path optimization and positioning, the coordinate system of the geometric representation is an important parameter with respect to run time. PSP supports different concepts for representing agents in 2D space. First, as an X and Y pair in a global frame 4.9a related to the digital map (GPS coordinates). Second, as a local, odometry-based coordinate system 4.9b using the vehicle's pose at start up as the center of the coordinate frame. Objects are mapped into this space using positions relative to the vehicle. This is valid position information for a short period of time. Third, as a pair of *S* (station) and *L* (lateral offset) in a *Frenet* frame 4.9c. Here, the current lane is used as a reference curve, which is used for all transformations into global coordinates.

Linking the space to the *global map* has been a common approach which uses high-precision GPS devices with additional correction data for position post-processing. However, the high costs of these systems as well as the need for external antennas drove the search for alternative localization methods. A lane markings-centric approach is one powerful alternative. Lateral localization can be done locally with a camera or laser system. A drawback of a lane-centric system is its missing link to the reference curve for lateral movements. Lane changes or strategic positioning in turns (e.g. minimizing curvature) is therefore complicated to express and needs additional coordinate transformations. Similarly, the complexity increases with an *odometry-based coordinate frame*, but a planned trajectory does not need to

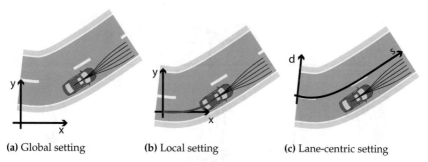

(a) Global setting   **(b)** Local setting   **(c)** Lane-centric setting

Figure 4.9: Three different types of coordinate frames for planning motions relative to a reference line.

be corrected along any reference. Position accuracy in odometry-based frames is transient, but stable for the time of a planning horizon for automated vehicles.

4.3 Visualizing high dimensional solutions

A major drawback of a universal planning system is its lack of driving mode aware-ness. We can observe a behavior from the outside and evaluate it, but the behavior itself was derived from many concurrent forces, which are hard to trace. In contrast to a mode of automation that deliberately chooses driving modes and organizes their execution, universal methods have a non-atomic decision approach. Although the universal planner is less traceable and less transparent, it can be measured. Its basis for all decision-making steps can be visualized, and therefore opened up for further investigation. How can we optimize results and solve issues in the devel-opment process? The manual supervision of a high-dimensional planning system with hundreds of thousands of partial solutions is impossible.

This is due to the massive number of states processed by the trajectory optimization. Visualizing hundreds of thousands of trajectories with a high level of similarity and cost implications is challenging.

A visualization tool is therefore created as part of PSP, called *PSPVis*. It is an ef-fective tool for debugging new planning solutions. Its primary use is to enable developers to investigate issues as they appear. To this end, the visualization uses PSP interfaces and acts on the three phases of sampling-based planning:

- Explore
- Generate
- Optimize

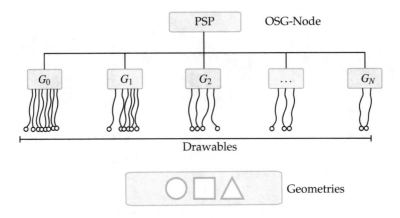

Figure 4.10: Modular visualization philosophy to display a high dimensional data set. The
search graph is displayed as a cloud of hundreds of thousands of 7D states.
Three dimensions are drawn as basic shapes (spheres, cubes or rectangles) in
different colors and levels of transparency.

It thereby, it differentiates between states, state transitions and cost trends. For each
phase a tool is present to investigate the most common issues.

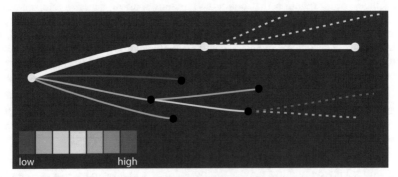

Figure 4.11: The search graph is displayed as spherical nodes and subnodes as well as
lines to indicate the affiliation. Larger spheres mark the beginning and the
end of state transitions. The colors represents the cost values (green = low, red
= high).

PSPVis implementation uses Qt as a graphical user interface (GUI) and *OpenScene-
Graph (OSG)* for 3D OpenGL visualization. It integrates well into the existing display architecture and acts as another layer on top of other visualization modules,

e.g. road representation or grid. As shown in Figure 4.10, PSP representations are stored as nodes of simple geometries in OSG display groups. Each drawable container is filled at each planning cycle regardless of whether its drawing option is active. We can thus pause the planner at any given time and manipulate our drawing settings for the particular snapshot. The visualization is as simple as possible. The geometries include points, lines, cubes, spheres and cylinders.

A debugging and control mechanism is present at each planning stage. This is shown in Figure 4.11. All samples drawn are visualized as points in a 3D space, where the X and Y axes are used to display position and the third axis Z is flexible, e.g. showing velocity or time. All transitions created in the trajectory generation phase are stored as a sequence of 20 states and each pair is linked with a line. The lines are colored according to the cost values to simplify the search for potential solution alternatives. All cost terms of every transition generated are stored and displayed in an extra Qt window. Each node has additional information on the actual cost for the last transition as well as the compound cost of the trajectory up to this node. Cost function evaluation is not straightforward and often has ambiguous results. By bringing simulation, planning system and search tree visualization together it is possible to trace the whole procedure for any snapshot and identify issues within the objective function. This tool has been used in all evaluation processes presented in this thesis. The details of the functionality are given in the experiment sections where necessary.

5 Sampling-based planning in phase space

The architecture of the PSP framework was introduced in Chapter 4. In this chapter, we want to focus on the implementations and solutions developed. It is begins with a detailed overview of a complete planning sequence in Section 5.1. this i s followed by the three planning stages of the PSP modules: State space exploration (Section 5.2), trajectory generation (Section 5.3) and optimization (Section 5.4). The trajectory generation is based on related work by Kelly and Nagy [62]. The optimization algorithms have all been implemented as CPU variants and some have been modified for use on modern GPGPUs [1]. Section 5.5 presents rules and heuristics for algorithm optimization.

Planning methods were chosen on the basis of the assessment criteria introduced in 2.4.2. This chapter addresses every criterion, except tactical maneuvers (see Chapter 6) and distortion (see Chapter 7). The framework is built such that it either complies with a criterion by design or it provides exchangeability of the component and interfaces for evaluation within the framework, itself.

5.1 Schematic of a complete planning sequence

Planning is a task that is repeated continually until a solution is found. All cycles are usually constructed in an identical way and a schematic for the PSP data flow and task list is presented in Fig. 5.1. There is one fixed time stamp named T_T that indicates the transmission of the last solution found to the control unit. This is done at the beginning of each planning cycle. Thus, the maximum planning time and frame rate are constant and other components can rely on them.

At the beginning of each optimization cycle T_P, the optimal trajectory with respect to the next transmission time T_T must be found. The remaining time is added to the scene and vehicle states as a state prediction. In the case of the ego vehicle, the last transmitted trajectory is used. If this does not exist, a state prediction can be made based on IMU and vehicle data.

At the end of the planning cycle the system idles for the remaining time T_{T+1}. If the system is under heavy load, this remaining time can be negative - and the point of transmission has been missed. This exception is handled by decreasing the planning frequency automatically, as long as a minimum frequency (5 Hz) can be guaranteed.

---

1 GPGPU - General Purpose Computation on Graphics Processing Unit

© Springer Fachmedien Wiesbaden GmbH, part of Springer Nature 2018
S. Heinrich, *Planning Universal On-Road Driving Strategies for Automated Vehicles*, AutoUni – Schriftenreihe 119,
https://doi.org/10.1007/978-3-658-21954-3_5

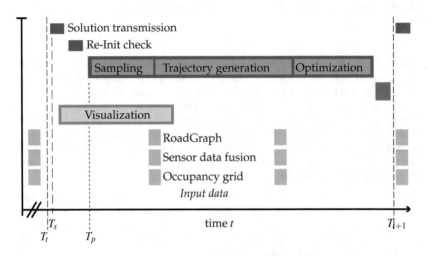

Figure 5.1: PSP schematic of data flow and order of events. T_T indicates the transmission
of the last solution to the control unit. Planning and visualization processes
are decoupled. The input data ($25Hz$) does not act as a trigger for the planning
instance. Planning is triggered independently with a fixed frequency, therefore
the next T_P is always known. Additional time is spent in idle mode (gray).

Check for a vehicle state re-init

So far it has been assumed that using a predicted point on the recently published
trajectory at time T_{T+1} will provide the true vehicle state. Unfortunately, errors
are accumulated within the motion control mechanism and subsequent actuator
control. This discrepancy between the ego vehicle's real state and the virtual state
must be dealt with within the planning world. When aligning the virtual state
with reality we call it a *hard re-initialization (re-init) process*. Whenever the starting
position is re-initialized, the control input suffers from discontinuities. This effect
can be smoothed out to a certain degree, but it is perceptible to the human driver
as a jerky motion (steering and accelerating).

5.2 State space setup and exploration

PSP creates a discretized representation of a phase space. It has been shown to be
effective to work in seven dimensions of the following intervals: $[t_0,t_1] \times [x_0,x_1] \times
[y_0,y_1] \times [\Theta_0,\Theta_1] \times [\kappa_0,\kappa_1] \times [v_0,v_1] \times [a_0,a_1]$, where time t, vehicle pose (x,y,Θ,κ), ve-
locity v and acceleration a are stored for a point mass in planning space.

The main axis chosen for planning is either time or space. Both approaches are addressed in this chapter. The coordinate system was a Frenet frame for the initial experiments and was changed to a local odometry frame in later stages of the evaluation.

The current lane is used as a reference in the phase of state exploration when new states are drawn and added to the search graph. All neighboring lanes are treated as alternative drivable regions and *regions of interest (ROI)*. A set of randomly drawn states is added at the end and rounds off the state exploration phase. The sampling strategy consists of one discretization parameter per dimension that distinguishes the total number of samples. An existing solution is always fed into the search space to allow stable decisions. In areas close to the ego vehicle a known state can be supported by a cost benefit. This is similar to inertia, which favors the old solution over similar solutions that require additional steering, for example.

Following the core idea of the PSP framework, the exploration interface allows multiple implementations and sampling strategies.

5.2.1 State propagation strategies

For planning horizons covering a large span of time (more than $10s$) or space, it is necessary to work with multiple temporal layers. Similar state propagation can be applied in other domains such as aviation [101]. Each state is propagated and the search graph grows exponentially. In contrast to simple end-to-end transition problems, the final target point loses its importance and is replaced with a target area. The solution density is relatively high in a state space with such a narrow workspace as occurs in automated driving.

An alternative to the state sampling method is a control or input sampling-based approach. In contrast to state sampling it supports the use of non-linear dynamic models and path and velocity sampling can be easily combined. For driving comfort systems a requirement towards smooth transitions (optimizing for least jerk) and path curvature is sufficient as long as the motion planning system does not request motions that push the system outside the circle of forces.

Expert knowledge supported sampling

It is helpful to draw samples based on expert knowledge if possible. In PSP a database of simple rules is implemented. Rules add waypoints (samples) that human expert drivers would use as well to reduce the complexity of the search space. Furthermore, knowledge on implemented cost terms allows waypoints in beneficial areas to be generated. This is used for cornering maneuvers and handling narrow passages. Both cases are known from digital maps.

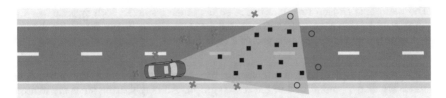

Figure 5.2: Samples are drawn from inside a cone-shaped zone. States that do not lie inside this zone and cannot be reached are shown in red. The blue squares are
samples inside the cone that represent feasible target states. States marked as
blue circles cannot be reached in the next sampling phase, but might be valid
later.

This also increases the density of drawn states in areas with information of high
value. An example is shown in Fig. 5.2. The cone-shaped area covers positions that
can feasibly be reached by clothoid paths. The cone spans both lanes and does not
include much waste (e.g. areas that are not drivable). A state that is not part of the
cone zone is marked in red (invalid). Others, marked as blue squares, are samples
inside the cone that represent feasible target states. States marked as blue circles are
out of reach in the current sampling step. They might be considered and be valid
in a subsequent iteration.

The result is far fewer invalid states in the search graph compared to a pure random
state sampling process. An improved ratio of valid states allows more efficient
optimization. Either more valid states can be compared or less time is spent in total
on finding the best candidate.

5.2.2 Vehicle motion guided sampling

In addition to expert knowledge, the motion constraints with respect to kinematic
or dynamic vehicle models are known as well. The complexity and cost of computation differ drastically, depending on the underlying model. However, in theory
any model can be linked to an exploration process. Dynamic models are treated
in a control sampling approach, where potential control inputs (e.g. steering angle,
acceleration) are simulated for a fixed time span. The resulting vehicle state is then
added to the search graph. A common approach is the kinematic or dynamic bicy-
cle model. For the latter, a linear tire model comprising a tire slip angle and a tire
cornering stiffness is assumed.

Control-based sampling has two major drawbacks in the use case under discussion
here. One is that it is difficult to express expert knowledge and therefore apply
search space simplifications. Secondly, it requires a good dynamic model, which
has high computational costs especially when time intervals between simulation
steps are required to be small in order to maintain precision and accuracy.

As this thesis addresses long planning horizons (up to 20s) and multiple exploration steps, the transition generation has to scale well. Invalid vehicle poses are ideally identified before solving for transition parameters. Thus, the cone shown in Fig. 5.2 guides the exploration. Its size is defined by vehicle motion constraints. The area of the cone is an underestimated prediction of positions that are reachable given the current vehicle state, control limits, course of the road and time.

5.2.3 Random state sampling

Algorithms such as R* and RRTs which use a randomized sampling process have been briefly discussed, as have their strengths and weaknesses. Even though they provide probabilistic completeness, the overall effort of finding a good sub-optimal solution in high-dimension spaces is challenging. However, randomized drawing of states in the sampling process is an effective method to generate options in addition to fixed sampling rules. Static rules do not serve well in cases where the ability to adapt is required.

5.3 Trajectory planning

The description of a vehicle's path over time is called its *trajectory*. It defines the characteristic and profile of an automated drive more than any other component of the self-driving system. For a period of several seconds the trajectory holds information about the future vehicle states (position, orientation, velocity and acceleration) at any given time. In the case of the PSP framework, the trajectory planning is divided into two components: Path and velocity profile generation. A trajectory has a temporal length of 10 to 20 seconds.

The kinematic or dynamic model defines the profiles of all driving behaviors. Simplifications can be linked directly to planning system limitations. A strict line is drawn between comfort driving maneuvers, and emergency or safety actions as well as racing. For this thesis, the focus lies solely on comfortable driving motions that do not push the vehicle towards the edge of the circle of forces. The trajectory generation methods implemented are described in detail in the following sections.

5.3.1 Generating path segments with clothoids

A feasible transition between initial and target state has to be found in order to connect two states. In path planning, a swerve-like motion can be expressed in several ways. The following procedure is based on a method that was presented by Kelly and Nagy [62]. Each path segment is defined as a third-order polynomial of the curvature κ over spatial stations s. It is defined by parameters $p = [a, b, c, d, s_g]$, where s_g is the path length and $\kappa_{p(s)} = a + b \cdot s + c \cdot s^2 + d \cdot s^3$. We want to find parameters $a - d$ as well as s_g for a transition that smoothly connects two states in state space.

McNaughton et al. [42] refine this approach by changing the parametrization to $p = [p_0, p_1, p_2, p_3, s_g]$ with the following constraints:

$$\kappa_p(0) = p_0 \tag{5.1}$$

$$\kappa_p(\frac{1}{3}s_g) = p_1 \tag{5.2}$$

$$\kappa_p(\frac{2}{3}s_g) = p_2 \tag{5.3}$$

$$\kappa_p(s_g) = p_3 \tag{5.4}$$

The idea is to align the order of magnitude for all parameters for the gradient descent. The function of curvature $\kappa(s)$ is therefore changed to:

$$\kappa_p(s) = a(p) + b(p) \cdot s + c(p) \cdot s^2 + d(p) \cdot s^3 \tag{5.5}$$

The newly introduced coefficients result in:

$$a(p) = p_0 \tag{5.6}$$

$$b(p) = -\frac{11p_0 - 18p_1 + 9p_2 - 2p_3}{2 \cdot s_g} \tag{5.7}$$

$$c(p) = 9 \cdot \frac{2p_0 - 5p_1 + 4p_2 - p_3}{2 \cdot s_g^2} \tag{5.8}$$

$$d(p) = -9 \cdot \frac{p_0 - 3p_1 + 3p_2 - p_3}{2 \cdot s_g^3} \tag{5.9}$$

In this case the third order polynomial guarantees continuity for the derivative of the orientation, the curvature. This allows smooth path descriptions. A higher order polynomial is desirable for the initial path segment in a multi-layer planning method. Thus, the derivative of the curvature rate can be used to ensure a smooth transition between two state transitions. Third order polynomials have two extrema as characteristic features resulting in up to three steering motions within one path segment. This gives us the following equations for the resulting position, orientation and curvature of the current goal state.

$$\theta(s_g) = \int_0^{s_g} \kappa(s)ds \tag{5.10}$$

$$y(s_g) = \int_0^{s_g} sin(\theta(s))ds \tag{5.11}$$

$$x(s_g) = \int_0^{s_g} cos(\theta(s))ds \tag{5.12}$$

The resulting path is determined by evaluating the polynomial spiral in the interval $[0, s_g]$. Equations 5.10-5.12 provide the missing state components. (x, y) is obtained by solving two Fresnel integrals, which do not have a closed form solution, unlike $\theta_p(s_g)$ and $\kappa_p(s_g)$. Using composite Simpson's rule, the integrals can be approximated as follows:

$$\int_a^b f(x)dx \approx \frac{h}{3}\left(f(x_0) + 2\sum_{j=1}^{\frac{n}{2}-1} f(x_{2j}) + 4\sum_{j=1}^{\frac{n}{2}} f(x_{2j-1}) + f(x_n) \right) \tag{5.13}$$

where $h = \frac{b-a}{n}$ and $x_j = a + jh$ for $j = [0, 1, \ldots, n-1, n]$. The interval $[a, b]$ is split into n subintervals, with n being a multiple of 2. Therefore, n directly influences run time and the accuracy of the algorithm. In this case n is set to 8.

A numerical approximation method is used for the path parameter optimization since some terms cannot be solved analytically. As previously mentioned, the parameters of the initial clothoid approximation are iteratively refined by using Newton's gradient descent method [102]. The spiral is parametrized with $p = [p_0, p_1, p_2, p_3, s_g]$ and p is optimized, such that the difference between the endpoint of the spiral $q_p(s_g) = [x_p(s_g), y_p(s_g), \theta_p(s_g), \kappa_p(s_g)]$ and q_{goal} is minimized and the spiral connects q_{init} and q_p.

$$z_p(0) = \begin{pmatrix} x_p(0) \\ y_p(0) \\ \theta_p(0) \\ \kappa_p(0) \end{pmatrix} \overset{!}{=} \begin{pmatrix} 0 \\ 0 \\ 0 \\ \kappa_0 \end{pmatrix} = q_{init} \qquad z_p(s_g) = \begin{pmatrix} x_p(s_g) \\ y_p(s_g) \\ \theta_p(s_g) \\ \kappa_p(s_g) \end{pmatrix} \overset{!}{=} \begin{pmatrix} \widehat{x_1} \\ \widehat{y_1} \\ \theta_1 \\ \kappa_1 \end{pmatrix} = q_{goal}$$

For convenience the function $z_p(s_g) = [x_p(s_g), y_p(s_g), \theta_p(s_g), \kappa_p(s_g)]^T \in C(\mathbb{R}^5, \mathbb{R}^4)$ is introduced[2]. As $\kappa_p(s_g) = p_3$ we can use $\widehat{z}_p = (x_p, y_p, \theta_p)$ in the following steps. The correction step is done using the Jacobian matrix $J = J_p(\widehat{z}_p(s_g))$ to calculate a correction vector, which is then applied to the previous set of parameters. The matrix is a 3×3 matrix due to the fact that entries can be neglected as $p_0 = \kappa_0$ and $p_3 = \kappa_1$.

---

2 s_g is explicitly given as a function input to avoid confusion as s_g is included in set p. Therefore, the expression can be simplified to z_p.

$$J_p(\widehat{z}_p(s_g)) = \begin{pmatrix} \frac{\partial x_p}{\partial p_1} & \frac{\partial x_p}{\partial p_2} & \frac{\partial x_p}{\partial s_g} \\ \frac{\partial y_p}{\partial p_1} & \frac{\partial y_p}{\partial p_2} & \frac{\partial y_p}{\partial s_g} \\ \frac{\partial \theta_p}{\partial p_1} & \frac{\partial \theta_p}{\partial p_2} & \frac{\partial \theta_p}{\partial s_g} \end{pmatrix} \tag{5.14}$$

The function $z_p(s)$ is evaluated with the current set of clothoid parameters. This is expressed as the distance error ε, which is essentially $\Delta\widehat{z}$.

$$\Delta z_i = q_{goal} - z_{p_i}(s_g) \tag{5.15}$$
$$\Delta\widehat{p}_i = J_{p_i}(\widehat{z}_{p_i}(s_g))^{-1}\Delta\widehat{z}_i \tag{5.16}$$
$$\widehat{p}_{i+1} = \widehat{p}_i + \vec{\alpha} \cdot \Delta\widehat{p} \tag{5.17}$$

Vector $\Delta\widehat{z} = (\Delta x, \Delta y, \Delta\theta)^T$ consists of the values of Δz without the curvature dimension κ. The learning rate α is a chosen value, which has an impact on convergence and step length of the gradient descent. When exit conditions are met, the algorithm returns its current clothoid parameter set $p_{(i+1)}$.

Look-up table for better initial parameters

The time to convergence of the gradient descent method depends greatly on the initial estimate of parameters p. A look-up table (LUT) can support and stabilize the optimization process as it stores parameters for an initial estimate. Instead of using default values, a LUT is used to find a better initial parameter set. A similar approach is presented by Knepper and Kelly [103] in the context of a lattice planning approach. A LUT returns the closest pre-computed parameter set p in the table given a query about relative differences between configurations q_{init} and q_{goal}. Therefore, the current state and the goal state are translated such that the initial state equals $[0, 0, 0, \kappa_i]^T$ and the terminal state is expressed as $[\widehat{x}_1, \widehat{y}_1, \widehat{\theta}_1, \kappa_1]$. The LUT stores these entries in a five dimensional grid, one dimension d_i for each variable. Each variable is uniformly sampled in a range between its limits $[L_{min}^{(d_i)}, L_{max}^{(d_i)}]$. For each grid point the corresponding set of parameters p_i is derived using a shooting method. A static mapping of indices as well as a general hash map implementation can be used to access the new parameters.

The attributes of the clothoid approximation itself help to find critical queries, such as a large lateral offset query at a relatively close station.

As the table is limited by memory space, a prioritization of initial parameter sets is recommended. Table 5.1 shows an example of entries for a path generation sup-

Table 5.1: Path generation: Look-up table for clothoid approximation

Index	p_0	p_1	p_2	p_3	s_g
$(\kappa_0, x_1, y_1, \theta, \kappa_1)$	...	...	...	...	...

Example of look-up table structure. All entries for p_i and s_g are parameters for the clothoid approximation. The search query requires initial and target values for the curvature κ as well as a relative position to the origin to shift the target state position (x_1, y_1) accordingly.

porting LUT. For most queries the LUT reduces the number of gradient descent iterations significantly.

5.3.2 Generating velocity profiles

Different time restrictions are added to the paths in a separate step. Although the path generation is independent of time restrictions, it is important to address vehicle-specific limitations. For example, path length, maximum vehicle acceleration and time might not fit together and form a valid expression.

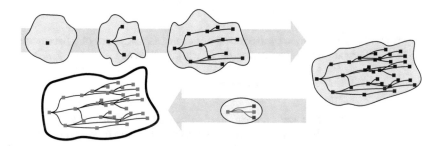

Figure 5.3: The feasible path configurations are used as seed pool for the velocity profile generation step. The search graph grows exponentially with the number of dimensions added to the velocity profile generator.

All path segments - state transitions without any dynamics - are fed into a velocity generator. The number of transitions increases correspondingly as shown in Fig. 5.3. Each path is described with its unique parameters p and an arc length s_g. The initial state sets the values for the velocity of the trajectory v_0 and acceleration a_0 at t_0. Similar to the idea of Xu et al. [45], the profile generated here allows for velocity and acceleration changes within a fixed time frame (in contrast to a fixed path length). Each velocity profile is expressed as a fourth order polynomial.

The terms for the six coefficients $g_0, \ldots, g_5$ are presented. The three functions $s(t)$, $v(t)$, $a(t)$ are defined as:

$$s(t) = g_0 + g_1 \cdot t + g_2 \cdot t^2 + g_3 \cdot t^3 + g_4 \cdot t^4 + g_5 \cdot t^5 \tag{5.18}$$

$$\dot{s}(t) = v(t) = g_1 + 2g_2 \cdot t + 3g_3 \cdot t^2 + 4g_4 \cdot t^3 + 5g_5 \cdot t^4 \tag{5.19}$$

$$\ddot{s}(t) = \dot{v}(t) = a(t) = 2g_2 + 6g_3 \cdot t + 12g_4 \cdot t^2 + 20g_5 \cdot t^3 \tag{5.20}$$

The velocity profile of the car should start with initial values for v_0, a_0 and travel a path of distance s_g in time τ. The car must have a terminal velocity of v_1 and acceleration of a_1.

$$
\begin{array}{ll}
s(0) = 0 & s(\tau) = s_g \qquad (5.21) \\
v(0) = v_0 & v(\tau) = v_1 \qquad (5.22) \\
a(0) = a_0 & a(\tau) = a_1 \qquad (5.23)
\end{array}
$$

The velocity profile generator has to find a continuous function of s, v, a that satisfies the constraints, such that the resulting transition connects q_0 (v_0, a_0) and q_{goal} (v_1, a_1) in time τ, where $\tau = \Delta t = t_1 - t_0$.

Additionally $v(t), a(t)$ should match the samples, which leads to 6 constraints:

$$s(0) = 0, v(0) = v_0, a(0) = a_0$$
$$s(\tau) = s_g, v(\tau) = v_1, a(\tau) = a_1$$

To satisfy all 6 constraints, we introduce 6 variables $g_0, \ldots g_5$. 3 out of 6 parameters can be set in advance, after evaluating the function at time t_0.

$$g_0 = 0, g_1 = v_0, g_2 = \frac{a_0}{2}$$

The remaining unknown variables can be obtained by solving the linear equation system. Given the values for $g_0, \ldots, g_5$, the velocity profile can be expressed for all $t \in [0, \tau]$ using the functions $s(t), v(t), a(t)$.

$$g_3 = \frac{20 \cdot s_g - 12 \cdot \tau \cdot v_0 - 8 \cdot \tau \cdot v_1 - 3 \cdot a_0 \cdot \tau^2 + a_1 \cdot \tau^2}{2 \cdot \tau^3} \tag{5.24}$$

$$g_4 = \frac{-30 \cdot s_g + 16 \cdot \tau \cdot v_0 + 14 \cdot \tau \cdot v_1 + 3 \cdot a_0 \cdot \tau^2 - 2 \cdot a_1 \cdot \tau^2}{2 \cdot \tau^4} \tag{5.25}$$

$$g_5 = \frac{12 \cdot s_g - 6 \cdot \tau \cdot v_0 - 6 \cdot \tau \cdot v1 - a_0 \cdot \tau^2 + a_1 \cdot \tau^2}{2 \cdot \tau^5} \tag{5.26}$$

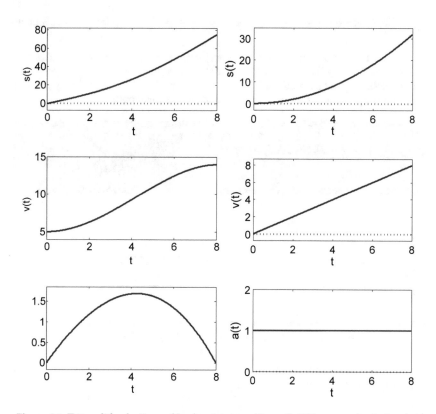

Figure 5.4: Two valid velocity profiles for state transitions: (left) Increase of velocity along the path and (right) constant acceleration.

By definition, the model is able to produce simple, constant acceleration profiles, as can be seen in (a), whereas different end constraints require a higher acceleration change rate (b). In the first example, the vehicle is constantly accelerated by $1\frac{m}{s^2}$ for 10 seconds. The second case shows a speed increase of $5\frac{m}{s}$.

A simple way to model acceleration for automated vehicles is to assume a constant acceleration value for each state transition. Although it is very fast to calculate, this simplification has several drawbacks:

- It is not possible to model continuous changes in acceleration

- Acceleration or deceleration is assumed to be constant for the entire length of the state transition. For long transitions (≥ 2 seconds) this leads to unusable profiles (e.g. braking maneuvers).

- Profiles based on constant acceleration are least likely to be described as *human-like*.

A comparison is shown in Fig. 5.5. Discontinuous acceleration profile queries will be smoothed out by the latency of the actuators and the drive-by-wire system. The actual profile cannot be predicted. The figure shows that different velocity profiles - methods or parametrizations - can be found for identical paths.

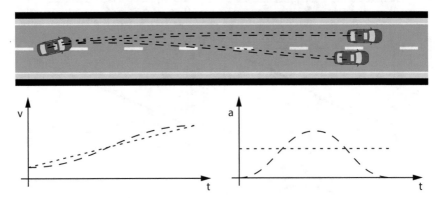

Figure 5.5: Different velocity profiles (e.g. a constant acceleration model) can be linked to identical paths by setting time limits accordingly.

5.4 Trajectory optimization

The discussion so far has dealt with how samples are drawn and state transitions are generated. To find the best sequence of state transitions via the state graph, a search algorithm has to be applied. A variety of search methods exist that optimize for the fastest response for a shortest path problem. To obtain a searchable graph structure, all transitions have to be collected during the sampling and generation processes.

5.4.1 Extended planning horizon with multiple layers

Single transitions are limited in their ability to describe complex motions. In contrast, a combination of swerve motions enables a sampling-based planning system to cover more driving modes. Actions and counter-actions can be expressed within one trajectory, such as accelerating and decelerating motions in the velocity profile or double lane changes.

Adding more planning layers creates a number of new challenges. First, the guarantees with respect to continuity do not hold in the waypoint state if no additional effort is made. Information that was present in the curve description is lost as it is not part of the state space dimensions. When using the waypoint state as the next initial state for the planning method, the trajectory might suffer from discontinuities at this point. Second, a problem that occurs when sampling states randomly is that some states are almost identical. Their separation in state space is small. This happens when the sampling is applied in state space rather than in the control space of a dynamic model. Grouping very similar states into clusters with one or more representatives reduces the number of waypoints and therefore increases the overall run time. Third, by using the same sampling strategies throughout all layers the number of nodes in the search graph grows exponentially with the number of layers.

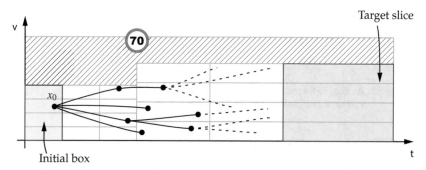

Figure 5.6: Planning in multiple layers along the main axis allows combinations of trajectories with different characteristics. (e.g. precise fusion of the acceleration and deceleration phase). A layer is called a planning *slice* and is in either the time or the spatial dimension. A receding horizon principle is used for the planning cycle. A temporary target region is selected for the current RoadGraph and routing information.

As shown in Fig. 5.6, the width of the cells increases with the distance in space and time towards the vehicle. At the planning horizon, the cells cover the largest state space area. This is due to the increasing uncertainty of the perceived environment and its predictability. A detailed plan at the end of the planning horizon is therefore not necessary. More important is an overall intention, e.g. executing a lane change or not.

Even though the horizon is extended by multiple layers, the global goal area is not covered for the majority of planning runs. A receding horizon principle is used for the planning cycle, whereas a temporary target region is selected with respect to the current road representation and routing information.

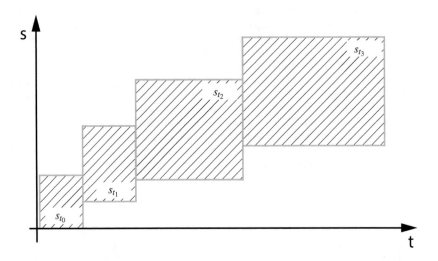

Figure 5.7: As time is the main axis for the proposed planner, the slices s_i (temporal intervals) have different sizes in spatial dimension.

Connecting path segments of multiple layers

Having multiple layers is beneficial as pointed out. What does this mean for the generation methods presented earlier? In PSP a layer is addressed as a *slice* of the planning horizon as shown in Fig. 5.7. This arrangement has the form of a multi-resolution structure, representing short-range and mid-range planning areas. Each slice holds a number of multi-dimensional containers (*boxes*) with a minimum of one representing a *box candidate*.

Slice The main axis is partitioned by a multi-resolution approach. A *slice* has a minimum length in that dimension (space or time) which increases with distance to the ego vehicle, forming short-range and mid-range planning areas. Slices hold a fixed number of *boxes* with the same length and volume in planning space, so that any transition in slice s_i ends in a slice s_{i+1} or greater.

Box A multi-dimensional representation for a container that represents an area in the discretized state space. Boxes are placed along a reference path and specifically exclude any part of $\mathscr{B}_i$. Boxes can vary in volume across different slices. All boxes hold information about any terminal candidates included.

Box candidate Each feasible state transition is considered to be a box candidate and assigned to a single container in the vicinity. In contrast to the representation of a state transition, a box candidate accumulates all costs from its parent candidates in the search graph.

First, a detailed overview of the search graph representation is given, as well as the mechanisms that are needed for a fast and reliable creation. Second, Dynamic Programming (DP) is introduced as optimization method and changes in the algorithm for CPU and GPU implementations of DP are presented.

Draw, compare, append, repeat.

As pointed out, path and velocity sampling are treated separately within PSP. All optimization methods are thus initialized with two instances, i.e. for sampling and for generation tasks. In Algorithm 1 this procedure is outlined in pseudocode.

Algorithm 1: Pseudocode of PSP trajectory optimization

Data: initial state x_i, target state x_f
Result: Π: the best trajectory candidate
createPSPStateSpace(x_i, x_f);
$slice = \text{rootInterval}(x_i)$;
while *exists(slice.next)* **do**
    $[x] \leftarrow$ pathSampling(slice);
    generatePaths($[x]$);
    $[\Pi] \leftarrow$ velocitySampling($[x]$, slice);
    generateVelocityProfile($[\Pi]$);
    $\Pi \leftarrow$ reconstructTrajectory($[\Pi]$);
if *invalid(Π)* **then**
    $\Pi \leftarrow$ emergencyStopTrajectory() ;
return Π;

The optimization method receives a current PSP state space based on the latest sensor measurements. The state space is directly partitioned into boxes along the main dimension (in this case time t), each segment on this axis is called a *temporal slice*. Each box $\mathcal{V}$ is initially represented by a candidate $\mathbf{q_i}$ with infinite costs in the center of the container. Other dimensions in the state space are the vehicle position (x, y) in Cartesian coordinates, orientation θ, curvature κ, velocity v and acceleration a. Each dimension has a different boxing granularity, whereby some are discretized more densely (e.g. position and heading values). The number of boxes per dimension is set as a parameter and held constant across all slices.

It becomes necessary to actively set *points of interest* to handle driving rules or road topologies which should start with a new slice or be held in the middle of a box. For example, a traffic sign indicating a change in the maximum speed is a position-dependent event. A uniformly partitioned set of slices might consist of a box that addresses a range of positions at a fixed point in time. Without adjusting the slice, the speed adjustment might be done close to the correct position but could be several meters off, depending on the length of the box. So, for important events, a slice will be divided into two independent slices.

All boxes of a slice have the same length with respect to the main axis. However, boxes do differ in expansion in other planning dimensions. Thus, the boxes do not have their center points at the same local position. Initially, each container stores a virtual box candidate with infinite costs. Boxes close to the vehicle's rear axis axle? are set to be small as the size increases over time in an attempt to allow high-resolution sampling for spontaneous motions and a lower resolution for trajectories at the end of the planning horizon.

A major feature for this thesis is a forward induction dynamic programming method. As the sampling process starts in the initial state it draws states into the neighboring slice (time difference of $1s$ for early slices).

The k-best cell representatives

As the graph search algorithms of choice explore the state space, newly drawn samples are likely to fall into identical cells. Thus, each cell selects k representatives and drops all other samples at the end of each exploration process within a slice. A representative is chosen by virtue of its accumulated costs over all layers, starting at the initial state. Only representatives are then used as initial states for further sampling in the next sampling iteration. All dropped samples are no longer part of the solution space. In addition to the guiding principle of the sampling procedure, the pruning adds to the drawback of returning a global non-optimal solution.

The size of k depends on the cell size and the overall discretization of the state space dimensions. In this thesis, the size of the cluster representative was set to $k = 1, 3$ or 5.

Whenever a sample falls into a box, it competes against the candidate with the minimal costs already in that box (*box representative*). Cost terms cover the distance travelled, jerk (acceleration rate) as well as proximity costs to a leading vehicle, thus allowing more than one candidate to generate subsequent states in the next sampling round. This algorithm is shown in Fig. 5.8. Candidates that did not make it into the group of the k-best box representative are skipped in the next sampling round. These candidates will become leaves in the overall search tree.

5.4.2 Directed graph optimization using dynamic programming

The creation of the search tree, in the exploration phase, results from a dynamic programming [12] algorithm with forward induction. It works well on such multi-level decision-making processes and relies on Bellman's *Principle of Optimality*, which is

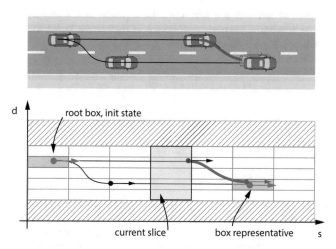

Figure 5.8: In each planning slice we fit containers (*boxes*) representing an interval in each
dimension. Samples drawn in the same boxes are evaluated by an objective
function and ranked accordingly. The k-best candidates represent the box.
Only these k states are used for the next sampling phase. The other path is
pruned to reduce the size of the search graph.

shown in Fig. 5.9. It is a combinatorial search in a discrete search space, which
guarantees a global optimal solution. A major drawback when using dynamic programming in high-dimension search spaces is that it suffers from the curse of the
dimensionality problem. The complexity increases exponentially with every extension of control or state space.

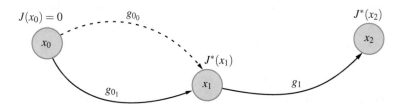

Figure 5.9: PSP related example of Bellman's *Principle of Optimality* for dynamic program-
ing.

Dynamic programming partitions a complex optimization problem into smaller
subsets. It is applied here to a PSP search graph and single state transitions. Let

Π be an optimal solution to the given planning problem within time interval $t = 0, 1, \ldots, T - 1$, where state x_i is present at time i. We now investigate the sub-problem in state x_i, which is to minimize the cost-to-go from step i until step $T - 1$:

$$J^*(x_i) = J_\Pi(x_i) = min \left\{ g_T(x_T) + \sum_{k=i}^{T-1} g_k(x_k, u_k) \right\} \tag{5.27}$$

The partition $\Pi_i \in \Pi$ is then the optimal solution to the sub-problem for time $k = 1, i + 1, \ldots, T - 1$.

Forward induction DP on a GPU

It could be argued that the problem presented is not large enough to make full use of modern GPGPUs. In a worst-case scenario, the implementation runs on the device with a low bandwidth and many expensive copy operations between host and device memory. Even though many samples are drawn and subtasks such as collision checking run concurrently, the algorithm cannot always be distributed well across the device so as to generate a steady, high throughput. Compared to state of the art computer vision algorithms, this seems like a fair argument. Nevertheless, a motion planning problem has (1) many tasks of a similar structure and size and (2) little interdependency between these queries.

The aforementioned planning algorithm could not be transferred into a CUDA version (NVIDIA's parallel computing architecture) without specific changes to its core:

- Each transition within the same slice is handled in a separate thread. This leads to a massive parallelization of the generation process.

- The representation of a transition is discretized into 20 states. A collision check for each state and approximated sub-states is performed in separate threads.

- Planning data (e.g. attributes of a state) is stored as arrays of structures to simplify access and avoid coalescing.

- All memory should be loaded and read in blocks that contain relevant information for a whole warp (32 threads).

- Ultimately, the solution (sequence of transitions) is written into the global storage. This includes optimizer meta data (e.g. costs, box affiliation) for debugging or visualization.

The trajectory generation method is implemented as a CUDA kernel. Frequently accessed data such as path parameters or costs is written to the device's shared memory. See Fig. 5.10 for a resource management description.

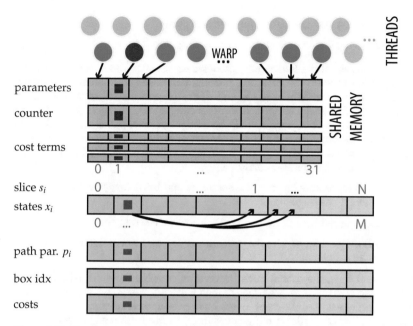

Figure 5.10: The state space representations have been optimized for GPU memory access. When the information is stored according to the *arrays of structs* principle, the kernel can load and write in blocks that are handled inside the same CUDA warp (32 threads).

GPU workload dependency on state space setup

For state space representations of sampling-based planners, the discretization of the main axis has a large impact on the effectiveness of massive parallelization. For each entry within a layer, several new samples are drawn and connected in the subsequent layer. On the one hand, the number of nodes grows exponentially, but on the other, it is not possible to calculate the results of subsequent layers concurrently, as tasks depend on results from previous layers. Time and spatial variants of the main axis in state space have already been discussed, although they share this particular problem.

In contrast to coordinate frames, time and spatial representations affect the workload differently. The dynamic programming approach has been developed as a Frenet frame, lane-based variant and also as an odometry frame, using local coordinates. The three options are as follows:

- Frenet frame with spatial representation

- Local odometry frame with spatial representation

- Local odometry frame with temporal representation

A Frenet frame simplifies the memory management as all road segments are treated as straight lines within the planning process, before the solution is transformed into global world coordinates at the end of the process. Road representation in the memory is drastically simplified. Furthermore, collision checks are easier in a Frenet frame as all vehicles are located on straight parallel paths. In an odometry frame, slices are positioned either in space or at certain timesteps relative to the vehicle's speed and the road's future topology, as slices, especially in turns, cover larger regions outside drivable corridors.

Although a Frenet frame has many advantages and leads to a simpler state space representation, it has huge limitations:

- Describing curved road or inner city scenarios with 90 degree turns.

- Calculating costs that require the curvature of the road and neighboring or intersecting road segments.

In the case of time, the spatial grids (interval of possible longitudinal and lateral reachable regions) are generated such that all reachable road segments are covered, whereas spatial grids of different layers have overlapping regions at low vehicle speeds as well as on roads with high curvature. In free form navigation especially, time is the superior description as positions can be reached multiple times in a trajectory.

In the case of time, the spatial grids (possible longitudinal and lateral reachable regions per time step) are generated such that all reachable road segments are covered. Whereas, spatial grids of different layers have overlapping regions at low vehicle speeds as well as on roads with high curvature. Especially, in free form navigation time is the superior description as positions can be reached multiple times in a trajectory.

For each time grid or spatial slice (depending on the main axis of the state space), threads are launched per state transition to calculate its costs. This triggers new kernels as each state of each transition is evaluated within one thread. The results are then written back as interim results. This procedure is illustrated in Fig.5.11. *CUDA Thrust* has been used within PSP to handle reduce and scan operations.

For PSP, it is more efficient by a factor of 2 to launch more threads and calculate a large number of invalid states in odometry-frame time-slices than to handle expensive global coordinate transformations within a kernel handling a Frenet frame representation.

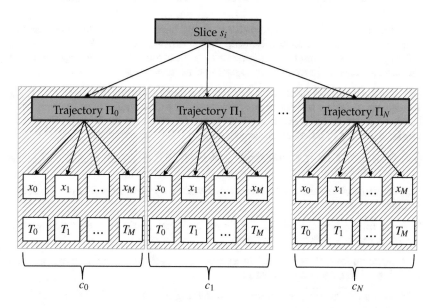

Figure 5.11: This Fgure illstrates the tree-like structure of CUDA kernels partition the workload for calculating costs within one planning slice. Each trajectory evaluated at single states along the trajectory. The final cost c_i is later accumulated.

5.4.3 Other PSP optimization methods

In addition to CPU and GPU variants of dynamic programming, randomized optimization methods such as R*GPU and RRT have been implemented as one option. While their strength may be that they outperform the chosen DP in less guided planning tasks, their inability by design to integrate problem-related heuristics has turned out to be a drawback. Especially in scenes with one to three drivable corridors, a randomized planner does not benefit from the fact that it is entirely free to pick from a state space which is far too large to process in detail. This argument does not hold when we add gear shifting to the state space and allow reversing as another maneuver.

5.5 Rules and heuristics

The methods developed in PSP do not possess any behavior state automation. The lack of such an extra layer in the planning hierarchy can be seen as a shift in control towards the layer of motion planning itself. The result is a flat hierarchy and the need for rules and heuristics in order to shape behaviors for automated driving. In contrast to single-purpose advanced driver assistant systems, the objective of a universal motion planner is to express many differing driving behaviors without the need for reparametrization or special initialization steps.

Hence, a basis of fundamental rules is required since planning as universal driving strategy provider has a broader perspective. Isaac Asimov's *The three laws of robotics* [104] and its enhanced version *The Four Laws of Robotic Cars* by Rojas [105] provide guidance for a rule-based behavior system of a self-driving vehicle:

1. A car may not injure a human being or, through inaction, allow a human being to come to harm.

2. A car must obey the traffic rules, except when they would conflict with the First Law.

3. A car must obey the orders given to it by human beings, except where such orders would conflict with the First or Second Laws.

4. A car must protect its own existence as long as such protection does not conflict with the First, Second or Third Laws.

Addressing the first two laws, universal driving behaviors in PSP use three independent heuristics:

Obstacle avoidance With respect to the first law, no valid motion plan is ever in collision with a static or a dynamic obstacle.

Lane keeping Although not regulated by federal law, it is advisable to drive within the lane boundaries. Driving in two lanes causes confusion and is not considered to be a safe maneuver.

Speed limit observance Traffic signs are an important piece of the traffic infrastructure. The speed limit is just one of a large set of signs that can be easily transferred into a heuristic for safer and more comfortable driving. The rules support traffic flow and benefit the safety of the parties involved.

As Asimov [104] pointed out rule-based systems have limitations for autonomous machines. It can always be shown that a conflict between all rules exist and a deadlock occurs that cannot be solved without breaking one of the laws. Solving unforeseeable problems in traffic is a difficult problem for automated vehicles and cannot be solved within a static set of rules. Humans use simple heuristics to decide on the next plausible move or answer [106], e.g. naming correctly the larger city out of a randomly chosen pair. Heuristics work well when the task context is comparable, for example as in on-road driving excluding crossing oncoming traffic. There are different groups of maneuvers, as mentioned in Section 2.4.2. Clusters 2, 5 and 6 are very different (e.g. crossing oncoming traffic) from basic maneuvering on roads.

A more detailed view on the cost function follows in Chapter 6.

5.5.1 Simplification of the state space

As mentioned before (5.4.2), optimization algorithms such as dynamic programming or RRTs suffer from the curse of dimensionality. As there is no Holy Grail for this problem, it comes down to increasing the computing power, a strategy of high parallelization, and drastic simplification of the state space as countermeasures.

Figure 5.12: Lane focused vehicle state sampling along the driving lane and the neighboring lanes (left) and control input sampling (right).

First, adding computing power is a good but limited approach, especially in the context of automotive electronic control units (ECUs). These platforms need to be energy efficient, light and compact, as well as powerful. In addition to these constraints, the platform for modern ADAS hosts more than one application, where the system resources have to be split between two or more parties. Second, a lane-focused state sampling process reduces waste (e.g. invalid and unfeasible states) in the search graph for structured environments, as shown in Fig. 5.12.

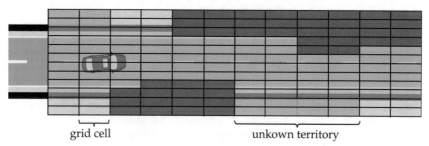

grid cell							unkown territory

Figure 5.13: Free space bands are attached to the reference line data structure. Each edge of a RoadGraph close to the vehicle holds free space bands. The band is a simplified representation of the occupancy grid. Free corridors (green), occupied space (red) or unkown territory (blue) can be described in free space bands. The size of a cell can be adjusted. The default size is 2×0.5m.

Third, we can deal with the complexity of sensor measurements and other input data. For the occupancy grid, we reduce the representation to a reference-path-centered free space in a pre-processing step, as shown in Fig. 5.13. This applies only to on-road driving tasks with precise in-lane localization. The assumption is that there exists at least one free-space corridor that is linked to a nearby driving lane.

Figure 5.14: Dynamic obstacles are predicted if their current position can be matched to a lane in the RoadGraph representation. Each object is predicted along the lane with respect to its current velocity and acceleration.

Finally, the prediction of dynamic traffic participants is stored and managed as follows: As dynamic obstacles are perceived in several sensors (e.g. radar, lidar and camera systems) and fused into a single representation, each vehicle observed is predicted for the length of the planning horizon (see next section for more details). Figure 5.14 illustrates how future poses of a vehicle are predicted along its driving lane. All poses are stored as a list with $250ms$ sampling time. A compact discretization of one single motion decision per dynamic road user is important as all predictions are tested against all transition candidates for collision.

5.5.2 How to choose a planning horizon length?

A question that remains is: *How much should we cover in our planning problem?* This relates to the outlook in time, space or driving behavior. In the latter, we need to take into account the execution time of a double lane change in contrast to a full braking task. As considering the whole route is not an option, a planning horizon has to be deliberately chosen. The parameters for the planning horizon t_H are as follows:

Vehicle speed The planning horizon has to be adjusted according to the vehicle's travel velocity. This does not affect the overall computation load as the search space resolution is constant. $t_H = min(t_{H_{min}} + 20 \cdot (\frac{v_c}{v_{max}}), 20)$

Road topology Exits, construction sites or road closures can all affect the future path. They might require one or more lane changes and therefore time for preparation is necessary. Fig. 5.15 provides more information on lane advise.

Maneuver length A double lane change or a combination of acceleration and deceleration within a single trajectory solution needs room in both the time and the space planning dimension.

As the vehicle travels, the reference lane may change due to the infrastructure or the road topology in general.

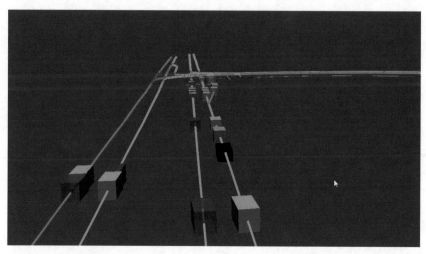

Figure 5.15: Between mission and motion planning, there is a strategic component called lane advise. Given the next required actions triggered by the infrastructure (e.g. highway exits), the module marks lane segments with a cost indication for driving on them.

6 A universal approach for driving strategies

All planners within PSP operate in one overall state (*drive*). The intention is to have no driving behavior state machine (or automaton) that guides the planning process on a higher level. Hence, every decision is made within the optimization process.

Section 6.1 describes objective functions of PSP as well as the kind of cost terms and their characteristics. It drives the decisions for an agent in a single state planner. These include handling merging traffic, circumventing other cars or responding intelligently to unexpected dynamic obstacles in general. However, parametrization of cost functions is a tough problem. The case where certain terms dominate others when their triggers are present (objects, road types) needs to be avoided by tuning the cost term itself and assigning weights. The optimization outcome is a driving strategy that has been shaped by criteria such as comfort or energy. Finding the ideal weightings to optimize driving behaviors was not part of the investigation.

In the second part of this chapter (Section 6.2), a method that supports and enhances perception performance for automated vehicles through smart positioning is presented. The planner receives a GPU-processed vehicle position cost map. Each entry represents the information gain for the system. This idea was published in [11] for IEEE Intelligent Vehicles Symposium in 2016.

In Section 6.3 the overall evaluation is presented.

6.1 Modeling driving behaviors: Cost functions

The basic idea of the PSP framework is to generate a large batch of trajectories and evaluate their quality concurrently in each planning slice. All state transitions are mapped onto a box structure. Each transition is then ranked against the current representatives in that box (cluster) on the basis of a cost function. State transition costs are independent of earlier transitions and are not affected by any other successive state. Furthermore, each cost component has to be as simple as possible in order to be fast to compute for all transitions of a slice.

The same principle can be reused for alternative planning approaches such as RRTs, R* or equivalent. Transitions are to be connected as long as they are collision free.

6.1.1 State transition costs

When two states are connected by a path segment and a velocity profile, the result is called a state transition. A planning solution Π_i consists of several successive state transitions $\Pi_i^{(k)}$ that connect the current state x_k and subsequent state x_{k+1}. The cost

© Springer Fachmedien Wiesbaden GmbH, part of Springer Nature 2018
S. Heinrich, *Planning Universal On-Road Driving Strategies for Automated Vehicles*, AutoUni – Schriftenreihe 119,
https://doi.org/10.1007/978-3-658-21954-3_6

function $\Omega : \Pi_i^{(k)} \to \mathbb{R}$ uses the cost components from Table 6.1 to rank the new transition $\Pi_i^{(k)}$ against the previous one. Therefore, Ω takes an input transition $\Pi_i^{(k)}$ and returns a numerical cost term c.

The total costs of the kth transition in a solution $\Pi_i^{(k)}$ are the sum of all weighted cost components. Each component is therefore assigned with an independend weight w_j.

$$c(\Pi_i) = \sum_k w_k \cdot c_k(\Pi_i^{(k)}) \tag{6.1}$$

The definition of each cost component is as follows:

Path length c_S The objective is to reach the target region as fast as possible while always meeting conditions such as comfort parameters. If the main planning axis is time, length becomes an indicator for fast trajectories.

$$c_S = s_g(\Pi_i^{(k)})$$

Lane keeping c_L This component supports trajectories that follow the center of the reference path r_0 with as little offset as possible. Thus, it helps the vehicle to stay in its lane and maximizes the distance to both neighboring lanes r_l and r_r or hard shoulders. c_L is dependent on the cost factor α which defines the lane keeping function's slope. A lane change is beneficial when the center of the other reference path is targeted. The width of the lanes are w_l for the left lane and w_r for the right lane segment. The offset γ outlines the need for at least another cost component to leave the lane (e.g. time).

$$d_j = \begin{cases} \alpha \cdot |\Delta r_0|, & \text{if } |\Delta r_0| \leq \frac{w_l}{2} \\ \alpha \cdot |\Delta r_0| + \gamma, & \text{if } \exists r_r \wedge \Delta r_0 < -\frac{w_l}{2} \\ \alpha \cdot |\Delta r_0| + \gamma, & \text{if } \exists r_l \wedge \Delta r_0 > \frac{w_l}{2} \\ \infty, & \text{otherwise} \end{cases}$$

$$c_L = \sum_j d_j$$

Curvature c_κ Driving in the center of the lane is not always beneficial. When taking turns a minimal curvature feels the most comfortable for passengers. Less steering is necessary in this case and this reduces the centrifugal force.

$$c_\kappa = \sum_j |\kappa_j|$$

Proximity c_P Adaptive Cruise Control (ACC) is one of the most popular DAS systems. Whenever a leading vehicle exists, c_p increases the total costs with decreasing distance to the vehicle in front. The parametrization was done on the basis of the ISO 15622:2010 standard [100], which requires a minimum time gap $t_{min} \geq 0.8s$. The preferred distance $d_p = 0.9s \cdot v_\Pi$ is therefore compared with the current distance d_c, when the time to collision (TTC) is less than $0.9s$. v_j stands for the planned velocity of the vehicle with respect to state x_j in transition $\Pi_i^{(k)}$.

$$p_j = \begin{cases} \infty, & \text{if } t_{TTC} \leq 0s \\ d_p - d_c, & \text{else if } t_{TTC} \leq 0.9s \\ 0, & \text{otherwise} \end{cases}$$

$$c_P = \sum_j p_j$$

Information gain c_E Some areas and objects in the car's environment deserve more attention than others (e.g. pedestrian crossing). The car can be forced to actively look out for such areas and observe them. But the sensor coverage depends greatly on the vehicle's position relative to other obstacles. A detailed description is given in Section 6.2.

$$c_E = \sum_j e_j$$

Acceleration c_A Ideally, the planner achieves all its targets with the least amount of acceleration, i. e. the velocity stays the same. Changing the vehicle's speed adds costs to the trajectory. For each substate, the calculation includes a term for the square of the acceleration as a term.

$$c_A = \sum_j a_j^2$$

Jerk c_J This component penalizes high rates of change in acceleration. As all planners are built as comfort systems, they should avoid high acceleration and deceleration within the same state transition, whenever possible.

$$c_J = \sum_j \dot{a}_j^2$$

Centripetal acceleration c_C As is the case with the curvature cost component, human drivers minimize centripetal forces by leaving the center of the lane in tight curves. This also helps to regulate the velocity selected in tight turns.

$$c_C = \sum_j |\kappa_j \cdot v_j|$$

Table 6.1: Terms of the objective function in the PSP framework

Cost term	Variable	Equation
Path length	c_S	s_g
Lane keeping	c_L	$\sum_j d_j$
Curvature	c_K	$\sum_j |\kappa_j|$
Proximity	c_P	$\sum_j p_j$
Information gain	c_E	$\sum_j e_j$
Acceleration	c_A	$\sum_j a_j^2$
Jerk	c_J	$\sum_j \dot{a}_j^2$
Centripetal acceleration	c_C	$\sum_j |\kappa_j \cdot v_j|$

Overview of cost terms that are used in a PSP objective function. To balance the terms and optimize overall behavior a weight is assigned to each term.

All state transition costs can be determined in situ immediately after generating a new state transition. The weightings w_i are chosen manually in an iterative process. In the simulation, the planning module is therefore confronted with scenarios (reproducibility) and each behavior cluster (see Section 2.4.2) is tested. As soon as a failure occurred, new parameters are set for the next run. This procedure could be automated in the future, whereby feedback could be bidirectionally communicated between simulation and planner.

Optimizing local solutions

A major drawback of this approach has led to research question 3 (see Section 1.2) of this thesis. Each driving strategy, regardless of its quality and sampling quantity , which is optimized by this set of costs may become invalid within a very short period of time in a real world scenario. One of the reasons for this is *motion uncertainty*. Modeling uncertainty for ego motion and choosing driving strategies with maximum durability is discussed in detail in Chapter 7.

However, a once optimal trajectory in a local planning problem can potentially become hazardous if unknown territory is treated as uncritical. As a special case, *occlusion* is a difficult problem for motion planning. Even though a planning system knows that information in perception data is missing or incomplete, it has to interpret how this should be dealt with. For scenes with many objects and prediction outcomes, the countermeasures vary greatly. As a result of this observation, a novel approach for automated cars is presented as part of this thesis. Similar to the philosophy of *active vision*, the vehicle is positioned such that information gain is optimized (see Section 6.2).

6.1.2 Optimizing for a universal driving experience

So how do cost components enable the planner to tackle very different driving modes simultaneously? There is no high-level planning instance and all scenarios are solved with exactly the same set of parameters and weightings, i.e. driving through narrow passages on a construction site is tackled with the same tools as a double lane-changing maneuver on a rural road. Rather than explicitly formulating what to do and which driving mode to execute, the costs describe the urgency with which things have to be avoided. In an ideal cost space there are many low-cost corridors that can be seen as valid and collision-free driving maneuvers along the available reference lines (lanes).

By adding a weighting to each cost term, it is possible to control the overall impact of a single component on the driving strategy. Basic behaviors such as lane keeping (distance to reference line), lane change (clothoid-shaped transition onto a new reference) and adaptive cruise control (proximity evaluation), can be seamlessly combined. As a result, each cost component has an effect on the timing and the general decision to leave the lane. Therefore, cost terms act as guidelines of unequal strength, responding with the most suitable driving strategy for the components chosen.

It is not possible to rely solely on costs and sampling strategy. The environment is cluttered with static and dynamic obstacles and some trajectories are in collision with one or more of them. Collision checks are done after the candidate's costs undercut one or more representatives of a box container.

Checking for collisions

Regardless of state space simplification and cost calculations, each transition has to be separately checked at the end of the box-matching process to ensure it is free of collisions. The collision check uses a very simple representation of the vehicle, similar to [69]. The ego vehicle is represented as 3 circles of similar radius as shown in Fig. 6.1. The algorithm operates as follows:

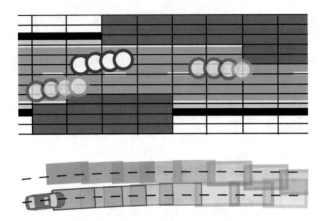

Figure 6.1: (Top) Collision check with static obstacles: A vehicle is represented as four circles. The check is performed against the free space bands - a RoadGraph annotated tri-grid structure. (Bottom) Collision check with dynamic obstacles: The
same representation is checked against every predicted dynamic obstacle close
to the vehicle.

1. If transition costs are lower than the kth best representative of the matched box,
   go to step 2. Else end here.

2. Perform collision check with static obstacles: Walk along the transition generated
   and check for collisions with the 3-circle model of the car, which has to be rotated
   according to the planned heading. If collisions occur, end here.

3. For dynamic collision checking, predict all other traffic participants that move on
   or across the current lane of the vehicle as well as vehicles in neighboring lanes.
   Compare positions and geometries between vehicle and obstacles at a sampling
   rate of 10Hz.

4. Return status: Free of collision

Combining elementary maneuvers

The distinction between simple and complex automated vehicle maneuvers has
been made in other publications [8] [42] [44]. However, widely established agree-
ment on the definitions of the terms does not yet exist. Reacting to merging vehicles,
circumventing other cars, or responding intelligently to unexpected dynamic obstacles have been identified as 'more complex maneuvers'. Maneuver combinations
allow more deliberative planning options. The optimized solution is more flexible and allows proactive driving decisions such as a sequence of accelerations and

decelerations to fit into a time gap. It has been found that a disadvantage of combining trajectories is that, as the complexity of the optimization problem increases, the run time slows down. In addition, it is worth mentioning that more layers and therefore combinations lead to an exponential growth of the search tree.

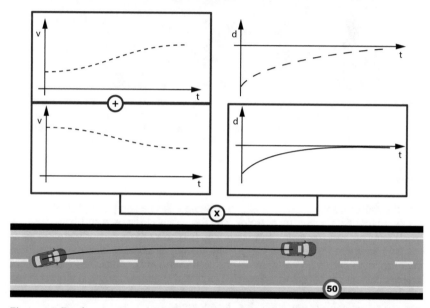

Figure 6.2: By planning in multiple layers and the decoupling of path and velocity profile generation it is possible to address more than two objectives in a single planning solution. As an example the planner chooses to brake until stopped while performing a lane change and reaching a time gap in a neighboring lane.

As shown in Fig. 6.2, maneuvers are created by stitching transitions from the multiple layer approach of PSP. As maneuver M_1 shows, an acceleration period M_2 reaches the time gap through a hard deceleration. The merging maneuver is described in a spatial dimension. The two PSP generators make use of the multiple layer structure of the search space to stop the vehicle immediately after reaching the neighboring lane while still reaching the necessary time gap - under the assumption that sufficient perception capabilities exist. Without any additional layers, the planning system must remain conservative with respect to its motions. Only a deliberate decision that takes all possibilities into account can provide more active behaviors.

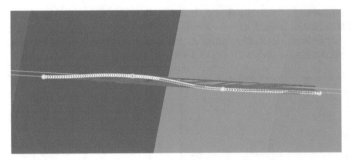

Figure 6.3: A double lane-changing maneuver viewed from above in PSPVis. Three slices (time) combine swerves to form the final maneuver. The initial swerve describes the first lane change, followed by another lane change back into the previous lane. The third segment allows to follow the reference path.

Example: Double lane-changing maneuver

An illustrative example is a double lane change. It is a complete plan of an over-taking maneuvers including merging into the neighboring lane and back after the overtaking process has been completed. Fig. 6.3 shows the box-candidate PSP planner in a simplified simulation setup. The maneuver consists of three parts. Two swerves into another lane and an acceleration phase in between. The white dotted line symbolizes the trajectory selected. The rest of the search graph is shown as red labeled lines. This was done as part of a simulator study. Enabling factors for such use cases are *Car2Car* communication technologies.

6.2 Situation awareness: Observing the most relevant things

Especially in urban areas self-driving vehicles suffer more disadvantages as human drivers are good at context understanding in cluttered scenes. Humans find it easy to communicate with other traffic participants by hand signals or verbal communication. In the event of occlusion, humans move their head to gain a better observation position. All these tasks are not straightforward for an automated car, but need to be addressed.

Therefore, the relationship between perception and motion planning modules has to be improved. As vehicle sensors are mounted at static positions on the vehicle, the only way for a car to reduce occlusion is to move. Perception modules themselves do not have any countermeasures against occlusion as their 360-degree sensor data fusion is also limited by obstacles blocking the view of the sensors. The

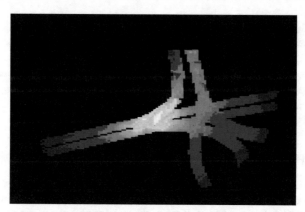

Figure 6.4: Representation of the entropy grid. The vehicle (blue arrow) is positioned in the top left lane. Grid cells are colored white to red to represent high to low entropy values. White regions are beneficial for the vehicle as it has the opportunity to gain further knowledge about its environment.

main idea is that there is always an optimal position or path that maximizes information about important areas. Similar to the philosophy of *active vision*, the vehicle is positioned such that information gain is optimized 6.4.

6.2.1 Optimizing the vehicle's sensor coverage

Instead of changing sensor mounting positions and orientations [89], the aim here is to compute promising positions with respect to their sensor coverage. In contrast to the vehicle's overall field of view (FOV), sensor coverage takes into account the fact that occluded areas potentially hold back valuable information. Thus, a vehicle's pose directly affects its sensor coverage, and information gain can be optimized through perception-orientated driving. A similar approach has been followed by Lidoris et al. [107] in the context of self-localization.

The idea is to positively influence the perception quality through motion planning and thus increase the planning performance in return. Therefore, measures for comfort systems like time-to-collision and distance to obstacles need to be extended. The approach presented creates a cost map that supports trajectories with an expected level of entropy. More information means a stronger basis for the next planning decisions. Therefore, it is not sufficient to optimize either the FOV or the sensor coverage. In fact, prior knowledge about important areas (highly dynamic or hazardous) along the route need to be collected a priori and annotated in the digital map. This process is done manually, but can be automated as the regions are to be found in the road graph. As an example, it is fairly easy to optimize for large sensor coverage that does not hold any useful information, whereas small sensor coverage with motion planning-related information can avoid a hazardous situation.

As human drivers do, it is easy to identify repetitive road topologies and situations that tend to be more risky than others. This is especially true for well-known stretches, such as daily commutes. The approach described here tries to reproduce a similar outcome for automated vehicles. The inputs for this module are road graph and occupancy data structures.

Building a sensor coverage cost map

The sensor setup for an automated vehicle is complex. It consists of redundant devices in case of any failure. It covers 360 degrees with measurements of different sensor technologies, such as camera, radar, lidar and ultrasonic systems. Each sensor type has unique capabilities and measurement properties (range, aperture angle) as well as qualities. As is the case with lidar-based occupancy grid, a cost map is created containing the entropy of poses (2D position and orientation) in the planning space. Fig. 6.4 shows a visualization of the final map. The map is later stored as a RoadGraph annotation and distributed to other modules such as the planning system.

Algorithm 2: Pseudocode entropy-based positioning

Data: Roadgraph rg, Grid W, Memory M, AOI A, DynObjList $objs$, Pose φ
Result: EntropyGrid E
$edge$ = findCurrentEgde(rg, φ);
$road$ = getEdgesAlongRoute(rg, $edge$);
while $exists(road.next)$ **do**
    while $pos = getLateralPosition(road)$ **do**
        while $h = getNextHeadingOffset()$ **do**
            W = mapDynObjectsInGrid(W,$objs$);
            M = updateMemory(M,W);
            idx = mapWorldToGrid($pose$);
            $E(idx,h)$ = entropy(idx,h,M,W,A);

return eg;

For this work, a simplification is made as the sensor field of view is deliberately underestimated. The representation chosen is an ellipse 25m long in the heading direction and 10m wide, as shown in Fig. 6.5. As is the case with ray tracing, the discretized ellipse consists of 100 rays, which is equivalent to 10cm gaps along the contour. The general sensing resolution is modeled with a precision of 10cm $\times$ 10cm, which is equivalent to the structure of the input occupancy grid. Algorithm 2 shows a brief pseudocode example of the procedure. The main effort can be clustered in three steps:

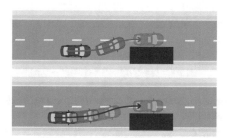

Figure 6.5: Hypothesis: Even for a mostly static scenery it is beneficial to calculate the chance of an information gain. This example shows that an earlier break point gives a better observation position to overtake the static obstacle (gray).

First, the current ego vehicle's world model is extracted from occupancy grid cells and stored in matrix W. Second, matrix M is used as the scene memory to indicate which area was observed in the past and whether information was available or not. Finally, all areas labeled as relevant are stored in matrix A. All matrices hold values of the interval $[0; 1]$ in each cell. For W and M the representation is as follows: Each cell describes an occupancy tri-state, where free cells are labeled as 0 and occupied ones as 1. If no information is available (areas not captured by the laser), the entry is 0.5.

This method is triggered by the RoadGraph module (25Hz). The cost map is built for each update, even though the actual planning frequency might be much lower. For each cycle it steps through the following sequence of tasks:

1. Diffuse the old scene memory M

2. Integrate sensor measurements from W into current memory M

3. Map dynamic objects into M

4. Explore future positions along lanes that are reachable from position p_0

5. Simulate sensor coverage for each pose (3D)

6. Compute the expected information gain

7. Build RoadGraph annotation

This module integrates well into the existing architecture. It does not depend on other modules as only inputs that were present before are used and the output is appended to one of them.

The scene memory M is either initialized with zeros or updated right at the beginning of the generation process. The update step diffuses the last scene and reduces the current knowledge about obstacles. The diffusion factor is 0.1 and has been chosen arbitrarily.

$$M[i]^{(t+1)} = M[i]^t - 0.1 \cdot (M[i]^t - 0.5)$$ (6.2)

The system has lost knowledge about the past state and is now ready to receive current sensor measurements. Therefore, matrix W is used to update matrix M. For this purpose each cell $M[i]$ is updated with the value of $W[i]$ if its absolute difference to 0.5 is smaller than the absolute difference between $W[i]$ and 0.5.

$$M[i] = W[i], \text{if abs}(M[i] - 0.5) < \text{abs}(W[i] - 0.5)$$ (6.3)

Now, the scene memory has not fully forgotten about the past and knows about the static world. By mapping all dynamic objects and their simplified geometries onto M coordinates, all information has been merged into one data structure. Cells of M that contain a dynamic object also have the value 1. The world model is now used to simulate future positions and the corresponding sensor coverage values. Therefore, the virtual vehicle is moved along the reference path and all reachable neighboring lanes. Sampling is done in three dimensions: Longitudinal, lateral and orientation. For each pose the sensor coverage area is simulated as an ellipse with a length of 25m and a width of 10m with its origin in the vehicle's center. As we step along a ray (see. Fig. 6.6), we need to map the points on the ray into the coordinates of M. The expected information gain $E[\varphi]$ at position φ is the Shannon entropy of the area covered in M. The entry is weighted by matrix A to indicate its relevance.

$$p(i) = M[i] \cdot log_2(M[i])$$ (6.4)
$$q(i) = (1 - M[i]) \cdot log_2(1 - M[i])$$ (6.5)
$$E[\varphi] = -\frac{1}{|C_\varphi|} \sum_{i \in C_\varphi} A[i] \cdot (p(i) + q(i))$$ (6.6)

C_φ represents the area covered at a simulated pose φ. Entropy is expected to be high when the vehicle covers areas that include areas of relevance and only sparse knowledge about the environment has been available.

The step size during this simulation is $0.5m$ in both the longitudinal and the lateral dimension. The heading is sampled from -20°to +20° in 7 discretized boxes using multiple resolutions: -20°, -10°, -5°, 0°, $\cdots$, +20°. The virtual vehicle moves on the matrix M and the final result is stored as a road graph annotation.

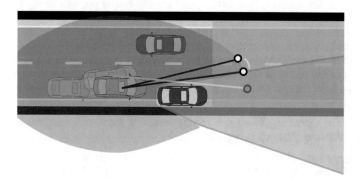

Figure 6.6: The representation of a 360-degree coverage based on fused sensor data measurements is an ellipse with 25m × 10m. This is massively underestimating the actual sensing capability in favor of faster computation times. The sampling resolution is at the outer bounds 10cm through 100 rays. The ellipse is used in a 3D workspace (2D position and orientation).

Optimization with CUDA

The bottleneck of the current implementation is in the process of generating the grid based on the sensor coverage approximation. Two major reasons for the CUDA implementation have been the requirement for a larger look-ahead distance in the grid and a shorter processing time. The problem is ideal for setting up a CUDA kernel. The operations are almost identical for each thread and the majority of memory operations uses entries that are stored close to each other.

By executing the operations concurrently the large number of matrix operations can be solved faster. Each ray is now calculated in a single thread and rays pointing into similar directions are grouped in the same CUDA warp. The change enabled PSP to process a grid of 40m instead of 30m and reduced the run time for one grid by a factor of 8.

6.2.2 Interpretation of a sensor coverage cost map

The objective is to present a cost map similar to the free space band (see Fig. 5.13) as a stand-alone road graph annotation. It will be linked to the corresponding edge of a lane. The road graph is one of the main inputs to the planner and is updated cyclically and shared across the platform. An example of an entropy grid E is shown in 6.4.

The vehicle's position is indicated by an icon in the upper left lane. The route foresees a right turn, crossing an area of interest (pedestrian crossing). The grid is colored from black (0) over red and yellow to white (1). As the simulation has been performed only at feasible positions, the grid entries form roughly the silhouette of the road topology. The white cluster is immediately after the vehicle's turn, facing the area of interest with no occluding obstacle in between. The tendency given by this map is to make a wide turn with higher than usual curvature (see other cost component c_κ).

The costs at a pose (x,y,θ) are derived by accessing the map through the states in the transition: $c_E(\Pi_j) = \sum_{x \in \Pi} w_E \cdot e_x$

On empty roads it is not necessary to update and maintain the memory. We define events that can trigger the generation of the information grid as follows:

- Geo event: Trigger stored in digital map.

- Road blockage: A dynamic obstacle occluding 50% or more of the lane ahead

- A braking target vehicle: Vehicle with a relative negative acceleration $\geq 1m/s^2$

So when activated, the entropy map is not only transferred into the cost space. The costs are accumulated for each trajectory and all the vehicle poses it describes. Therefore, other components can be built upon this term, e.g. to control vehicle speed. Accumulating costs for driving through areas with high information gain would generally only show results in spatial planning dimensions (swerving).

6.3 Simulation experiments

This section discusses whether a universal system can produce results comparable to those of specialized planners and how local knowledge can be improved. All tests were applied in *Virtual Test Drive (VTD)*, a simulation suite of VIRES, extended by Volkswagen Group Research. VTD allows test loops with different levels of abstraction. The most lightweight setup for developers generates basic inputs such as occupancy grid, dynamic object list and road graph from VTD's world model.

Validation setup and hardware

Even though simulation is a closed and strictly controlled process, there is a wide range of planning parameters and hardware configurations. To make this process transparent we need to outline the setup in detail before discussing its outcomes. Reproducibility and determinism cannot be guaranteed in real world closed- loop testing (vehicle feedback). System latencies, sensor measurement noise characteristics and on-board hardware impact the outcome of a control problem. For the experiments described here, sensor data quality is assumed to be ideal.

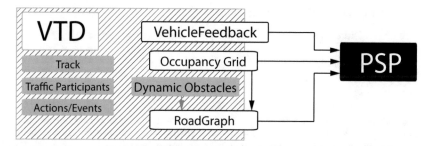

Figure 6.7: Lightweight simulation setup for motion planning module developers that generates basic inputs such as occupancy grid, dynamic object list and road graph from VTD's world model.

A detailed diagram of the data provided by the simulation is shown in Fig. 6.7. VTD has an internal world model, which is used to generate all data structures that are known and relevant to the automated driving system. The road topology is extracted and published cyclically. It includes all annotated information, such as free space bands, speed limits, vehicle localization match or traffic light states. Most importantly, the vehicle dynamics are calculated on the basis of the latest trajectory input.

All tests have been conducted on a machine with a 2.7GHz i7-4800MQ processor and 16GB memory. For CUDA support, a NVidia Quadro K2100M device with 576 CUDA cores and 2GB GDDR internal memory was chosen. It is important to mention that the entire setup runs on this unit. That includes a virtual machine with the simulation suite, as well as the ADTF communication framework and PSPVis modules for visualization.

PSP configuration and parameters

In order to produce comparable results, the configuration of PSP is kept static throughout all driving mode validation tests. The setup chosen is shown in Table 6.2. The PSP library is used by a local planning implementation with an odometry coordinate frame and *time* as the main planning axis. All modules in use have different data triggers and timings. Table 6.3 lists the main providers and subscribers of PSP representations and inputs with the communication properties.

The planning horizon depends on the time dimension and is discretized in 7 slices with a multi-resolution approach, having more slices closer to the vehicle. The planning frequency is 5 Hz on average, as stated in Table 6.3. PSP itself is cyclically self-triggered and reads the latest input data from RoadGraph, grid and simulation. For more details see Fig. 5.1 in Section 5.1.

Table 6.2: An example of a PSP configuration

Dimension	unit	range (min)	range (max)	intervals
time t	s	0	20	5
orientation θ	$°$	$-\pi$	π	5
curvature κ	$\frac{1}{m}$	-1	1	10
velocity v	$\frac{m}{s}$	0	40	20
acceleration a	$\frac{m}{s^2}$	-3	2	3

Detailed PSP experiment configuration. The state space is sliced along the time axis covering position (depending on current state and predictions), orientation and curvature as path parameters. Velocity and acceleration with respect to the main axis form the velocity profile of the trajectory.

Table 6.3: ADTF filters and data rates

Module	frequency
Simulation	40 Hz
PSP	5 Hz
RoadGraph	25 Hz @ 180 kb/s
Grid	25 Hz @ 150kb/s

Overview of executing frequency of data providing modules and planning module.

6.3.1 Validation of driving strategy generation

The objective of all driving experiments is to validate whether the PSP algorithm is capable of generating different driving modes based on its cost function. Furthermore, different modes are to be combined seamlessly to form new driving strategies. This proof of concept follows the driving clusters introduced in Section 2.4.2. In the following paragraphs, each driving mode cluster and its objective function term as well as potential events that trigger the outcome will be discussed.

Start, follow and proximity scenario

In the case of start, follow as well as proximity driving modes, the overall maneuver share similar characteristics: For all maneuvers, the ego vehicle drives in the same lane, and it may or may not have a leading vehicle. All three scenario sequences are illustrated in Fig. 6.8. As in the case with a starting maneuver there are reasons why the vehicle was stopped (e.g. traffic jam, traffic light or road blockage). Traffic assumptions are very different, depending on the road type and the reason for the standstill: At a traffic light, we can assume that as soon as we get a green light, all vehicles try to reach the permitted speed limit. In stop-and-go traffic the initial pose is similar to the one at a traffic light, although the acceleration behavior is very

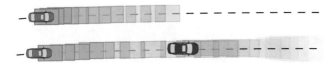

Figure 6.8: (Top) Simple starting behavior without a lead vehicle. (Bottom) Following behavior with a lead vehicle. Similar to adaptive cruise control, the follow and proximity behavior is shaped by the prediction of the lead vehicle.

different. Why is this important? If leading vehicles are predicted to be accelerating more slowly than their real counterpart, the ego vehicle will plan its trajectories with the limits of the virtual obstacle. The process of following is driven by the *proximity* and *path length* cost term. As the vehicle tries to drive as far as possible it is restricted by the leading vehicle. Following has the particular challenge of keeping the distance constant as dynamic obstacles change their speed or momentum. It is important to ensure that the sampling process always draws the predicted motion of the leading vehicle. Otherwise, the system is forced to weaken the current following state and indirectly increase costs in later planning cycles. The proximity behavior allows a positive velocity difference of up to 0.9s TTC.

One problem identified is that the traffic prediction quality needs better situation awareness. Otherwise, a planner such as PSP will always over- or underestimate acceleration profiles of other dynamic objects and therefore open unintended gaps to the leading vehicle when starting to drive, for example.

Double lane change scenario

A double lane change is a composition of three or more state transitions. However, the controlled swerve in X/Y-space to a neighboring lane and back is not the only dimension that can benefit from such a construct. Similarly, we can now form acceleration and deceleration (or vice versa) sequences to get into a critical time gap (e.g. for overtaking or merging). In the scene shown in Fig. 6.9, the car executes a double lane change and is fully aware of when and where to merge back into its driving lane. Such a maneuver can include several lane changes over up to three lanes if sufficient perception information is provided.

A still unsolved task is achieving similar results with real world data as sensor measurement noise weakens the ability to predict the future traffic state. This becomes even harder for predictions of oncoming vehicles. First, they are picked up late by the sensors as they usually drive in an occluded area. Second, they are far away and approach quickly. And third, as lane merger has uncertainties itself, the lane mapping is a hard task for the vehicles ahead.

Figure 6.9: A double lane change requires stable and precise perception measurements. The maneuver ensures safe states throughout the maneuver until the vehicle reached the center of the initial lane.

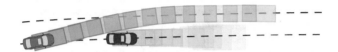

Figure 6.10: A trajectory to change lanes in dynamic traffic, which is predicted for the length of the planning horizon.

Lane change scenario

The classic lane change is simpler than the previous behavioral cluster. Although, it is a concatenated sequence as the double lane change, it does not require as much perception data and foresight to ensure no collisions. So from a planning and control point of view, the results are redundant. The illustration of the maneuver is shown in Fig. 6.10. However, the maneuver can be extended by a more complex velocity profile, in this case an accelerating profile that ensures an earlier time gap.

A drawback of the single state approach is that, especially for lane changes, the process of changing lanes is not stored internally and therefore cannot be published across the system. Without further adjustments, this prevents the indicators being set well ahead of the maneuver.

Taking turns

One reason to change the internal reference frame to global or semi-global was the problem of taking turns in *Frenet frames*. As the reference curve is added at the end of the generation process, the road appears straight throughout the whole planning cycle. For curves with a small curvature value this can be ignored, but 90-degree turns on urban roads need a different steering behavior than driving always in the center of the lane. Cutting curves and therefore minimizing the curvature κ along the trajectory can be straightforwardly addressed in the objective function.

Figure 6.11: Illustration of a planned trajectory for driving through turns. This behavior is optimized for minimum curvature.

As shown in Fig. 6.11, the vehicle leaves the reference curve and slightly overshoots the path at the end of the turn. Overall, the new track has a lower curvature value.

As these examples show, a universal planning module does not need a high-level behavior state automaton. With respect to research question 1 it has been possible to show that the system complexity can be reduced to a single state planner generating driving strategies over multiple layers. Maintenance and parametrization are shifted from several behavior modules into one general objective function. Different behaviors can be combined and executed simultaneously or as a deliberate planning sequence.

6.3.2 Evaluation of smart positioning

Robot local knowledge based on on-board sensor measurements is the most trustworthy source for an automated vehicle platform (except for human driver interventions). Compared to backend or V2X solutions, vehicle perception is always up-to-date and has a relatively short error chain. Parts of this evaluation have been published by Heinrich et al. [11].

In this evaluation it can be shown that smart positioning is a way to increase local knowledge. Thus, both motion planning and perception benefit from this feedback loop between the two modules. First, a scenario where three independent trajectories are ranked by the optimization process is considered. Whereas the normal planner would optimize for proximity and lane keeping, the new cost term shifts the algorithm's attention towards lateral offsets to increase the situation awareness. In a second scenario the environment is more dynamic. Time is important to observe relevant areas and the planner has options such as stopping or acceleration to collect more information. The assessment involves the observation and interpretation of two values. The experiments should examine if the earlier hypothesis holds, i.e. that the largest possible sensor coverage is not the overall best criteria.

Area of interest entropy (AIE) A measure to express local knowledge of the automated vehicle on relevant areas at a waypoint. The lower this value, the more information can be gained by adjusting the trajectory such that it includes the waypoint.

Field of view coverage of AOI (FVCAI) A measure to describe the ratio of relevant and irrelevant areas covered by the vehicle's FOV. The higher this value, the more information rich areas are covered.

The initial situation is similar for both cases. The automated vehicle arrives at a three-way stop. The relevant areas have been marked and stored in RoadGraph annotation.

Scenario 1: Lateral positioning

The distance between the automated vehicle and the leading vehicle is 18m. The distance at standstill is set to be 2m. To simplify the selection process, there are only three trajectory candidates which are distinguished only by their lateral shifts (0.5, 0, -0.5), as shown in Fig. 6.12. The adjacent lane in front of the leading vehicle is marked as an area of interest. Oncoming traffic from both directions would be expected at this intersection.

For each trajectory, the entropy grid is calculated and AIE as well as FVCAI are distinguished and shown as plots in Fig. 6.13a and Fig. 6.13b. For all trajectories there is a decrease of 0.25m along the initial 6m. The AIE then decreases with a steeper negative slope and reaches its minimum at roughly 7.5m. At a position close to the leading vehicle (2m) the AIE indicates different knowledge gains ($\Pi_1 : 0.6, \Pi_2 : 0.68$). For Π_3 we get the smallest value and therefore the highest information gain for the adjusted vehicle position in this scenario. Similar behavior can be observed for the FVCAI. In order to maximize the sensor coverage of the relevant areas, the vehicle has to leave the center of the lane and adjust its view.

Scenario 2: Velocity profile adjustments

The initial position of the automated vehicle is 7m behind a static obstacle. The task is to plan a swerve and pass the obstacle in the adjacent lane before the stop line is reached. Here, the relevant areas are around the crosswalk (right) and the oncoming traffic (left). The vehicle's path is shown in Fig. 6.12.

The vehicle's sensor coverage of relevant areas is initially still beneficial. This is shown in Fig. 6.14 with a low AIE value (0.2). There is a rapid increase as soon as the vehicle moves towards the obstacle, which is due to the increasing occlusion. As a result the crosswalk is now out of sight. This cannot be changed until the sensors pick up the hidden areas again, which happens right at the intersection stop line. From here on the AIE decreases even further until it reaches almost zero. Everything has been observed and is well known to the system. This result indicates that an expected gain in robot local knowledge can be used to cause a velocity reduction. Unexpected changes in the environment need adjustments in the driving

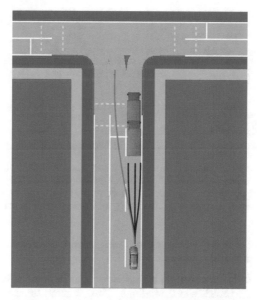

Figure 6.12: A VTD screenshot including an overlay of potential paths for two scenarios is shown in this image. The black trajectories illustrate three options for scenario 1. Lateral offset within the lane can add benefit to the vehicle's environment model. The red trajectory describes a path of scenario 2. Depending on the position, the vehicle can observe relevant areas and react with an improved velocity profile.

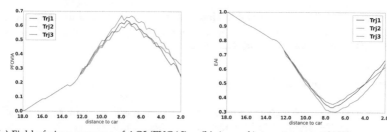

(a) Field of view coverage of AOI (FVCAI)   **(b)** Area of interest entropy (AIE)

Figure 6.13: The two plots cover AIE and FVCAI values for smart positioning scenario 1. For all three evaluated trajectories there is a decrease for 0.25m along the stretch of 6m. The AIE then decreases with a steeper negative slope and reaches its minimum at roughly 7.5m. At a position close to the leading vehicle (2m) the AIE indicates different knowledge gains (Π_1 : 0.6, Π_2 : 0.68). For Π_3 we get the smallest value and therefore the highest information gain. Similar behavior can be observed for the FVCAI.

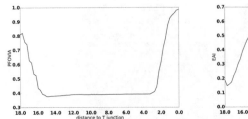
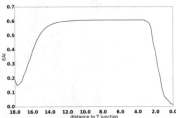

(a) Field of view coverage of AOI (FVCAI)   **(b)** Area of interest entropy (AIE)

Figure 6.14: The vehicle's sensor coverage of relevant areas is initially still beneficial. The low AIE value (0.2) highlights that. The two plots cover AIE and FVCAI for smart positioning scenario 2. As soon as the vehicle moves towards the obstacle there is a major negative change in AIE due to the increasing occlusion.

strategy and entering any of these areas with a large relative difference in speed leads to an unpredictable outcome.

Both cases show that the idea of active vision can be applied to strengthen the knowledge on the vehicle's surroundings. Thus, it contributes to the broad local knowledge about the vehicle's surroundings. This knowledge of unfiltered information gathered by on-board sensors can be used to form decisions on driving strategies.

6.3.3 PSP performance analysis

The performance tests are twofold. First, the impact of the GPU implementation and the adjustments to the CUDA specific algorithms are investigated. Second, the hotspots in the general PSP workload are discussed.

Planning runtime analysis

The transition to a fully GPU accelerated algorithm has been done incrementally. Initially, main hotspots of PSP workload have been implemented as CUDA kernels. This hybrid approach soon reached its limits when memory management prevented any performance benefits. Therefore, the whole PSP planning core was implemented as CUDA modules. In this evaluation all three variants will be covered and the implementations details are as follows:

- CPU: Implementation of path and velocity generator using hyper-threading.

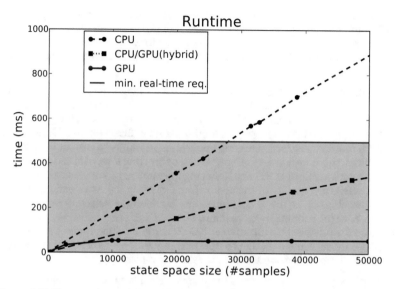

Figure 6.15: Three variants of the same planning system have been implemented. (1) CPU with hyper-threading support, (2) a hybrid CPU/GPU version with several operations between host and device memory, and (3) a full GPU implementation. The minimum baseline for a real-time planning system is assumed to be 500ms.

- CPU/GPU hybrid: Exploration is done on the CPU per slice. Paths are generated in a CUDA kernel and pushed back to the host system. Velocity sampling is done by the CPU, followed by a kernel call for velocity profile calculation. Solutions are stitched and wrote back into new data structures on the CPU. Frequent memory copying between host and device hinders a better overall throughput rate.

- GPU: This variant forces changes to be made to the PSP interfaces between exploration, generation and optimization phases, as all of them are executed on the GPU device. Thus, the results are written back to the host system, but more information is lost compared to the other variants.

The configuration shown in Table 6.2 was used as a base in this comparison. In this series of tests the actual driving outcome is less regarded. Therefore, a static scenario was build up to evaluate each implementation similar to unit tests. Hence, simulation, visualization or ADTF communication was excluded as all modules ran as part of a lean structure without any overhead.

As an example, throughout, one planning cycle 60000 trajectories are generated with 500 unique initial positions. In our tests $10-15\%$ of all trajectories do not land in a valid region and are used to prune the search tree. In this setup an average of 80 state candidates land in a box, from which the top 3 act as representatives. The rest is discarded.

The results presented in Fig. 6.15 show that, even though the GPU bandwidth is not fully used, the run time can be improved significantly compared to the other implementations that involve a CPU. The CPU uses only hyper-threading to provide a concurrent generation process. As a minimal goal, 2Hz was considered as the minimum requirement for a real-time application in traffic. In case of the CPU implementation this criterion was not met when optimizing a set of 50000 trajectories, which required a little over 900 ms. Reducing the number of samples or the number of slices, effects the quality of the result. Implementing the workload hotspots as CUDA kernels, such as the calculation of the Jacobian for trajectory optimization, help to reduce the runtime by more than a factor of 2. However, it was not possible to increase the number of samples over 100000 without falling under a rate of 2Hz. The GPU-only implementation in comparison ran with over 10Hz and scaled well in the same setup with respect to the number of samples.

A valid proposal for an increased frequency that holds for all implementations is to decrease the number of slices. Without sacrificing too much accuracy, 10-12 spatial slices and 5 temporal slices produced reasonable results. As trajectory generation is a repetitive action of similar tasks with respect to arithmetical operations and processing load. It is therefore ideal for massive parallelization on a GPU as a CUDA kernel. The hybrid version suffers from long copying times between host system and GPU device in combination with short processing times for the batch of samples on the device itself. The GPU variant allows real-time applications of the planner, whereas CPU and the hybrid version must be scaled down with respect to sampling density in order to run with 2Hz or more.

Workload hotspot analysis

PSP cycles perform a variety of tasks with different complexity. While some tasks need large amount of data loaded and stored temporarily, others spend time on sole computing problems or idle due to dependencies with other tasks before terminating. All three cases can be identified as a hotspot, where the algorithm spends a majority of its processing time.

In order to analyze all potential bottlenecks and inefficiencies the toolbox *Intel VTune Amplifier XE 2013* was used. Whenever a tasks is identified as a hotspot it is also a the best target to gain performance improvements. For all three implementations, the hotspots (time spent on processing the task) vary. This is due to the level of concurrency or hardware architectures (e.g. number of floating point units). For example, as shown in Fig. 6.16 the Jacobian operations of the gradient

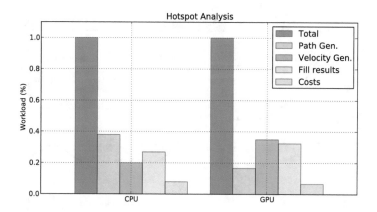

Figure 6.16: Comparison of CPU and GPU workload hotspots. Executing the path generation threads highly concurrently adds the biggest value to the run time optimization. The GPU workload is defined by the high effort of sampling velocity profiles and gathering the result information and store it correctly in global memory.

descent for the trajectory generation tasks dominate the analysis of the CPU processing workload. Over 50% of time is spend on this procedure. The rest is used to invert the matrix and compute the end state of the trajectory based on the optimized set of parameters. An optimization of the Jacobian expression reduced this hotspot by 14%. Therefore, the initial number of operations in the expression was decreased by a factor of 5. Further optimization for the CPU implementation was achieved by adding path integration and a LUT for sine and cosine functions.

In comparison, the GPU implantation hotspots were different. For all measurements the *NVidia Visual Profiler* is used as it provides detailed insights and statistics about each kernel. First, the kernel with the highest run time is identified with the profiler. Second, idle times are investigated first. Third, different kernel runs are maximized as they run asynchronously and copying memory is postponed as much as possible. Fourth, it has shown to be worth to except additional processing load within kernels when this increases the overall concurrency.

One major hotspot of the GPU implementation is the process of filling state containers in order to create solutions along the planning horizon. This is mainly a memory managing task at the very end of the main algorithm, which does not allow much parallelization.

Similar to the CPU variant, one drawback is the heavy use of sine and cosine operations. Although, this can be compensated by using look-up tables instead of CUDA sine or cosine implementations, it still remains a factor in the final result.

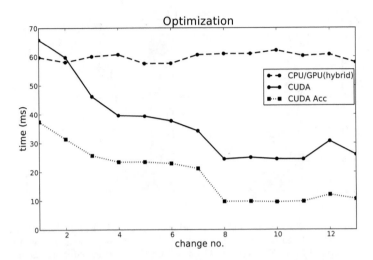

Figure 6.17: Profile of applied changes to either GPU kernel code or general PSP representations. Measures that have proven to optimize run time on the GPU. All
changes did not negatively effect the CPU implementation and partially add
benefits to the CPU implementation.

The following list includes measures that have proven to optimize run time on the
GPU:

- Avoid divergence: Threads within the same warp should execute the same code.

- Runtime-API calls are expensive: Avoid latencies of these calls or combine them
  with required kernel calls. Same applies for memory operations. Reuse allocated
  memory if useful.

- Slices in PSP are expensive: No runtime-API call within a slice-loop.

- Introduce cached allocator: Reuse allocated memory.

- Maximize for concurrency: Trade additional processing for more concurrency.
  Example: *DevStateAt* is a method to retrieve a state based on its predecessor. The
  new *StateAt* method uses integration over 8 sampling points, but is therefore
  independent. The additional load is 8 times higher, but we get 40 times more
  concurrency and achieve a speedup of a factor of 5.

- Optimize for block occupancy and therefore fitting base block sizes (change no.
  2 in Fig. 6.17).

- Rearrange order of planning dimension for faster accessibility.

- Reduce collisions of atomicAdd operations.

- CUDA Thrust usage: Avoid calling cudaGetDeviceProperties multiple times in one session (change no. 3 in Fig. 6.17).

Figure 6.17 shows a run time profile after applying the aforementioned changes. Changes, such as number 8 an improved cost calculation structure (as shown in this chapter), improve the algorithm run time by a factor of 2. Other methods include change of data types (using short integers for indexing and sorting), cached allocators or reorder of data types stored in a PSP representation memory.

7 Modeling ego motion uncertainty

In Chapters 4 to 6, the full planning stack was introduced and evaluated. In this chapter, a post-processing method is presented which compensates for ego motion uncertainty. So far, it has been assumed that the optimized trajectory would be executed by the vehicle's control and actuator systems with no or only minor deviations. The trajectory actually driven can differ far more than that, however. Even though the planning results are still completely safe, these circumstances would cause frequent re-initializations. But any re-init strategy (presented in Section 5.1) is only a countermeasure and no solution to this problem. The idea of handling ego motion uncertainty within the planner was published in [9] for IEEE Intelligent Robots and Systems (IROS) in 2015. This chapter goes beyond the initial proposal.

7.1 Why modeling uncertainty matters

A trajectory actually driven may differ from the planned trajectory. The planner presented reacts with a re-init at the beginning of planning cycles. This process resets the initial pose from an earlier planning result to the current localization pose. Position and orientation experience a sudden reset with the consequence of a jerky input on steering wheel or throttle. In addition to a localization problem as shown in Fig. 7.1b, ego motion uncertainty addresses future vehicle poses and their dependencies by means of a vehicle dynamics model.

In addition, a receding horizon approach has the problem of not having global knowledge, which is present in all local planning approaches. For example, it is straightforward to think of a trajectory that allows an automated car to merge behind another vehicle. Let us assume vehicle ahead was predicted along its lane on the basis of a high acceleration measurement. For one, acceleration of other vehicles is hard to measure and it is even harder to predict is its future behavior. So for an prediction in the ideal world, the time gap would be a fit for a merge behind the car. The optimizer chooses the extreme maneuver on the assumption that every option is known in the decision process. As time passes, the system could find out that there is not as much space as initially assumed. Thus, all alternatives plans are then invalid. Ideas on how to cope with this issue in the context of automated driving have already been presented (e.g. hysteresis) [42]. Likewise, the measurements of the ego motion can be erroneous as well. Terrain, slope, weather or inertia can pull planned and real-world execution apart.

Modeling vehicle state uncertainty works as well for all traffic participants and their motion predictions. The linear-quadratic Gaussian (LQG) method that is the basis for the approach used in this thesis was also used in [108] to model the vehicle's

© Springer Fachmedien Wiesbaden GmbH, part of Springer Nature 2018
S. Heinrich, *Planning Universal On-Road Driving Strategies for Automated Vehicles*, AutoUni – Schriftenreihe 119,
https://doi.org/10.1007/978-3-658-21954-3_7

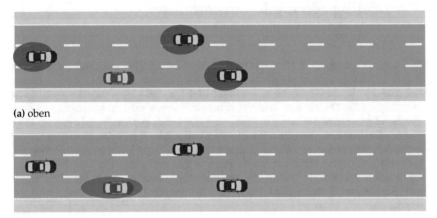

(a) oben

(b) unten

Figure 7.1: All measurements are noisy and have errors. Perception sensor measurements describe object properties, such as size and motion. Localization methods and motion sensors measure the current state of the ego vehicle. Errors depend on various parameters: Ego vehicle speed, orientation change rate, sensor characteristics and quality, dirt and others.

environment. The LQG approach supports the planning system by canceling out potential trajectory candidates that are likely to be invalidated after a small number of control cycles. Fewer re-init calls add value to the robustness, which is addressed in research question three.

Similar to probabilistic sampling-based planners, the post-processing method has the property of being *probabilistically complete* [13], as the algorithm's probability of solving the given (solvable) problem converges to 1 when the run time approaches infinity. Thus, it is guaranteed that the post-processing method will find a solution to the trajectory generation problem, provided that it has a sufficient amount of time.

7.2 Ego motion uncertainty

Every real-world control problem has to deal with uncertainty. This also applies to vehicle driving modes that do not touch the limits of the circle of friction. Terrain properties, slope, wind or latencies in the vehicle's communication and drivetrain systems can cause increasing numbers of measurement errors in the control loop, as shown in Fig. 7.2a. As it is not possible to eliminate the cause, it can be compensated for by modeling the existence of motion uncertainty into the system. The state transition is described by a function which contains the vehicle's dynamics and a

(a) A feasible, collision-free trajectory describing a lane change. It's temporal validity is dependent on the trajectory's curvature and velocity profile as well as the distance to the predicted positions of other traffic participants.

(b) The ego motion uncertainty grows with increasing numbers of steering actions within a driving behavior and depending on the velocity profile.

Figure 7.2: Ego motion uncertainty differs for two related, valid planning solutions.

predetermined control vector $\mathbf{u_k}$ in addition to the effect of a disturbance variable $\varepsilon_k = \left(\varepsilon_\kappa^{(k)}, \varepsilon_a^{(k)} \right)$ at stage k for curvature and acceleration, which is drawn from a zero-mean Gaussian distribution with variance M_k:

$$x_{k+1} = f(x_k, u_k, \varepsilon_k), \ \varepsilon_k \sim N_n(0, M_k). \tag{7.1}$$

As a result, the states also become random variables through the effect of stochastic quantities. Longer-lasting trajectories are identified by using this well-understood LQG method [109] to estimate the likelihood of future collisions paying due consideration to the vehicle's motion uncertainty. Consequently, this post-processing method contributes to an increase in the temporal validity of the trajectory selected.

7.2.1 A post-processing planning extension

The approach presented in this chapter builds upon the work of van den Berg et al. [109]. They presented a LQG motion planning system where a linear-quadratic regulator (LQR) and a Kalman filter for state estimation form a LQG control system. It is used to model the motion uncertainty of a variety of different systems. The authors applied this method to a simple, non-holonomic vehicle problem using a RRT planner for trajectory optimization. In this work, the method of Berg et al. is extended and used as a post-processing method of PSP trajectories or other motion planners of the Volkswagen Group Research department.

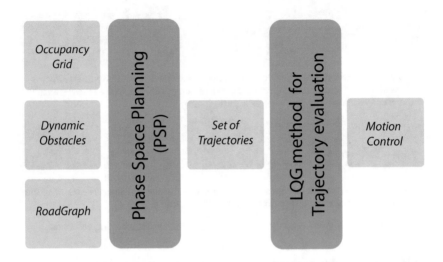

Figure 7.3: The LQG post processing method has the following data flow: (1) A planner
generates an ordered set of potential solutions, (2) each trajectory is evaluated
by the LQG method, (3) trajectories that are likely to interact with objects in the
near future are removed from the set and (4) the solution with minimal costs is
returned.

In contrast to the original data flow shown in Fig. 7.3, the framework returns a set
of N planning solutions per planning cycle as a series of states $(\mathbf{x_k})_{\mathbf{k=0,...,l}}$, where
$N \geq 10$. The set is ordered according to the costs to reach the target area at the planning horizon. Each trajectory Π is a set of pairs of temporal uniformly discretized
($\Delta t = const$) states $\mathbf{x_n} = (x_n, y_n, \theta_n, v_n, t_n)$ and controls $\mathbf{u_n} = \left(u_\kappa^{(n)}, u_a^{(n)}\right)$, where $n \in \mathbb{N}$.
The control input has two components. The curvature or orientation rate κ, and
vehicle's acceleration value a. Therefore $\Pi = (x_0^*, u_0^*, \ldots, x_{l-1}^*, u_{l-1}^*, x_l^*)$, and it follows
that

$$x_n^* = f(x_{n-1}^*, u_{n-1}^*, 0) \tag{7.2}$$

In Chapter 5 of this thesis, a velocity profile generation method was introduced
and combined with a clothoid-based path planning system. The vehicle dynamics
based on this approach are of the form $x_{n+1} = f(x_n, u_n, \varepsilon_n)$ with:

Figure 7.4: Inputs to the post processing step are results from a previous planning cycle. The set of planning results includes many driving strategies. The algorithm shall identify a solution that is long lasting with respect to the interference of other traffic participants. All trajectories come with an existing cost evaluation (lowest: green, highest: red).

$$x_{n+1} = f_1(x_n, u_n, \varepsilon_n) = x_n + \cos(\theta_n)v_n\Delta t \tag{7.3}$$
$$y_{n+1} = f_2(x_n, u_n, \varepsilon_n) = y_n + \sin(\theta_n)v_n\Delta t \tag{7.4}$$
$$\theta_{n+1} = f_3(x_n, u_n, \varepsilon_n) = (u_\kappa^{(n)} + \varepsilon_\kappa^{(n)})v_n\Delta t \tag{7.5}$$
$$v_{n+1} = f_4(x_n, u_n, \varepsilon_n) = v_n + (u_a^{(n)} + \varepsilon_a^{(n)})\Delta t \tag{7.6}$$
$$t_{n+1} = f_5(x_n, u_n, \varepsilon_n) = t_n + \Delta t. \tag{7.7}$$

As we go along the initial trajectory states x^* and compare a value at step k to the newly generated state x_k, we keep the difference $\bar{x}_k$. The same procedure is applied to the control values:

$$\bar{x}_k := x_k - x_k^* \tag{7.8}$$
$$\bar{u}_k := u_k - u_k^* \tag{7.9}$$

In addition to a vehicle model, it is necessary to describe the confidence in a single perception measurement update. For a vehicle in state x_k at time k, a function for the current sensor measurement update h provides a new update with an assumed disturbance ξ_k:

$$z_k = h(x_k, \xi_k), \ \xi_k \sim N_n(0, N_k).$$

Sensor data $\mathbf{z_k}$ is described in five dimensions for each time step $k \in \mathbb{N}$, and noise is simply added to the state x_k by means of ξ_k. The sensor observation equation results in

$$z_k = h(x_k, \xi_k) = I_5 \cdot x_k + I_5 \cdot \xi_k$$

The estimate of the sensor measurement is compared to the function h given the
input state at step k and no further noise.

$$\bar{z}_k := z_k - h(x_k^*, 0) \tag{7.10}$$

The idea of re-init procedures in the case of large deviations between measured
and planned vehicle states was introduced in Chapter 4. The control system keeps
the vehicle close to the path at all times. Therefore, an approximating a non-linear
model by a Taylor expansion along the input trajectory Π gives us the linearized
stochastic dynamics and the observation model:

$$\bar{x}_k = \quad A_k\bar{x}_{k-1} + B_k\bar{u}_{k-1} + V_k\varepsilon_k, \ \varepsilon_k \sim N_n(0, M_k) \tag{7.11}$$

$$\bar{z}_k = \quad H_k\bar{x}_k + W_k\xi_k, \ \xi_k \sim N_n(0, N_k) \tag{7.12}$$

where A_k, B_k, V_k, H_k and W_k are Jacobian matrices defined as follows:

$$A_k = \frac{\partial f}{\partial x}(x_{k-1}^*, u_{k-1}^*, 0)$$

$$B_k = \frac{\partial f}{\partial u}(x_{k-1}^*, u_{k-1}^*, 0)$$

$$V_k = \frac{\partial f}{\partial \varepsilon}(x_{k-1}^*, u_{k-1}^*, 0)$$

$$H_k = \frac{\partial h}{\partial x}(x_k^*, 0)$$

$$W_k = \frac{\partial h}{\partial \xi}(x_k^*, 0)$$

This is a common formulation of LQG control [110]. As an optimal control is to
be calculated using this formulation, the cost function is defined by two symmetric
positive-definite matrices C and D. While C holds the state variables, the matrix D
holds the penalty terms for the chosen control inputs.

$$g(\bar{x}_k, \bar{u}_k) = \sum_{k=0}^{l} \bar{x}_k^T C\bar{x}_k + \bar{u}_k^T D\bar{u}_k \tag{7.13}$$

The cost function in Equation 7.13 is minimal when $\bar{u}_k = L_{k+1}\bar{x}_k$, where the feedback
matrix L_k is derived by a standard finite-horizon discrete-time LQG controller [110].
Thus, Equation 7.11 can be simplified to

$$\bar{x}_k = (A_k + B_kL_k)\bar{x}_{k-1} + V_k\varepsilon_k. \tag{7.14}$$

Consequently, this can be rearranged by defining $\zeta_k := V_k \varepsilon_k$ and $T_k := A_k + B_k L_k$. Then $\tilde{x}_k$ can be written as $T_k \tilde{x}_{k-1} + \zeta_k$, which is a common form of the Kalman filter expression. A problem arises since $\tilde{x}_k$, the vehicle's state difference at step k between trajectory and motion model, is unknown to the system. A vehicle state $\tilde{x}_k$ has to be obtained. As the form of $\tilde{x}_k$ suggests, a Kalman filter is used for the state estimation.

$$\bar{u}_k = L_{k+1} \tilde{x}_k \tag{7.15}$$

7.2.2 State estimation and model correction

A Kalman filter follows a simple paradigm of a continuous sequence of prediction and correction steps, whereby its internal estimator is adjusted in each step. The overall goal is to estimate the minimum mean square error for a state $\tilde{x}_k$ as an affine function of the observations $\bar{z}_1, \ldots, \bar{z}_k$. This is the same as minimizing the following function:

$$\mathbb{E}(||\tilde{x}_k - \tilde{x}_{k|k}||^2)$$
$$= tr\left[\mathbb{E}((\tilde{x}_k - \tilde{x}_{k|k})(\tilde{x}_k - \tilde{x}_{k|k})^T)\right].$$

where $tr(A)$ is the trace of a quadratic matrix, i.e. the sum of its diagonal entries.

A Kalman filter cycle has two main phases: *Prediction* and *correction*. The phases work as follows:

Prediction A Kalman filter uses its internal estimator for a *prediction* step, even though measurement z_k is not present at this stage. Instead previous observations including $\bar{z}_{k-1}$ are used. This estimator is defined as $\tilde{x}_{k|k-1}$. The estimator depends on its predecessor $\tilde{x}_{k-1|k-1}$ and the dynamic model:

$$\tilde{x}_{k|k-1} = A_k \tilde{x}_{k-1|k-1} + B_k \bar{u}_k. \tag{7.16}$$

Correction In a second step, new measurements $\bar{z}_k$ are used to improve future estimations. This part is called *correction*. A Kalman filter has a gain matrix for this purpose. The gain matrix modifies the estimator and the covariance matrix of the last prediction step:

$$K_k = P_{k|k-1} H_k^T \left(H_k P_{k|k-1} H_k^T + W_k N_k W_k^T\right)^{-1}. \tag{7.17}$$

Here, in simplified terms the correction ratio between the current estimator and the latest measurement is determined to form a new estimator. Therefore, the covariance of the a priori approximation error $P_{k|k-1}$ is multiplied with the Jacobian

H_k. H_k contains the partial derivative of the sensor function h with respect to the state variable x. The partial derivatives of the sensor function h with respect to the sensor noise variable ξ_k form the Jacobian W_k.

$$\tilde{x}_{k|k} = \tilde{x}_{k|k-1} + K_k \left(\bar{z}_k - H_k \tilde{x}_{k|k-1} \right). \tag{7.18}$$

7.2.3 Re-evaluate trajectory collision probability

Collision checking is a major part of PSP trajectory optimization. Trajectories that are evaluated in this post-processing step have been checked in the PSP optimization process and proven to be collision free. With the ego motion uncertainty at hand, we have the possibility to re-evaluate the best trajectory candidates with respect to their probability of colliding with an obstacle. Hence, a more geometrically complex procedure is needed to check for collisions between a position uncertainty along a trajectory and an object in the scene. In this thesis, a *Gilbert-Johnson-Keerthi (GJK)* algorithm is applied. GJK requires all obstacles to have a convex shape and thus it will divide up each obstacle polygon. This property can be used directly for the collision estimation:

- For each time step k, the state space is linearly transformed using the matrix U_k^{-1}. This transforms the position uncertainty ellipse into a unit circle.

- The unit circle scaling factor c_k is obtained by calculating the Euclidean distance between the circle center and the closest point of all surrounding obstacles. c_k is the largest scaling factor for the unit circle (obtained from the covariance ellipse transformation), so that the scaled unit circle B_1 does not intersect with any obstacle geometry.

- The chi-squared distribution is now expressed as a regulated gamma distribution:
  $P : \mathbb{R}^2 \to \mathbb{R}, P(x,y) = \frac{\gamma(x,y)}{\Gamma(x)}$, where $\gamma(x,y)$ is the incomplete gamma function. For each stochastic variable $X_k \sim N_n(\mu_k, \Sigma_k)$, $k \in \mathbb{N}$ and c_k scaled unit circle B_{c_k}

$$P\left(X_k \in B_{c_k}(\mu_k, \Sigma_k) \right) = P\left(\frac{n}{2}, \frac{c_k^2}{2} \right). \tag{7.19}$$

- For simplification, it is assumed that all terms are independent, when calculating the total probability of avoiding any collision along the trajectory. The total probability is:

$$P\left(X_k \in C_{free}, \ \forall \, k \in \mathbb{N} \right) \approx \prod_{k=0}^{l} P\left(\frac{n}{2}, \frac{c_k^2}{2} \right). \tag{7.20}$$

- The best of all N input trajectories with the maximum value from Equation 7.20 is chosen. This trajectory is passed on to the systems control unit.

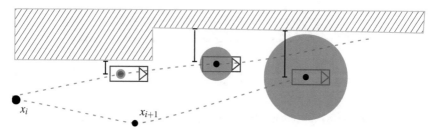

x_i

x_{i+1}

Figure 7.5: Unit circle scaling factor c_k is the largest scaling factor for a unit circle B_1, such that B_1 does not intersect with any obstacle geometry. Left: A small covariance ellipse allows for a larger scaling factor. In this case, the path segment investigated is considered to be safe. Center: For a larger but still feasible position uncertainty c_k is decreased. This example is considered to be less safe, expressed by a smaller c_k. Right: Scaling factor c_k for a trajectory with a high degree of position uncertainty.

Unit circle scaling factor c_k

The factor c_k is defined as the largest scaling factor for a unit circle such that the resulting circle B_1 does not intersect with any obstacle geometry. This procedure is shown in Fig. 7.5 As long as the random variable falls within the scaled ellipse there is no collision. The probability is defined as:

$$P(X \in B_c(\mu, \Sigma)) = F_{\chi_n^2}(c^2), \qquad (7.21)$$

where F_{χ^2} is the chi-squared distribution with n degrees of freedom.

In this definition, the fact that the vehicle representation is a simple point mass is neglected. Any uncertainty in orientation would affect the obstacle polygons that have been enlarged by the vehicle's geometry assuming constant orientation.

Let us discuss three cases of collision estimation with respect to different amounts of uncertainty: High accuracy, small position uncertainty, large deficit in position localization. We assume for this comparison that, without loss of generality, the ellipse is a unit circle and its the transformation U^{-1} its identity matrix I.

1. For small covariance ellipses as shown by way of example in Fig. 7.5 (left), it is unlikely to collide with the static obstacle. This can be expressed with a value of $c_k > 3$. The path segment investigated is considered to be safe.

2. For a larger but still feasible position uncertainty, c_k is decreased. In comparison the example is considered to be less safe for a $c_k < 3$ as shown in Fig. 7.5 (center).

3. For small c_k values as shown in 7.5 (right vehicle), e.g. $c_k < 1$, the distance between vehicle and obstacle is either small or the position covariance ellipse large.

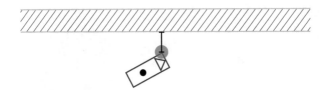

Figure 7.6: Correct interpretation of the factor c_k at the closest point to the obstacle.

For increasing c_k, the collision probability decreases continuously. This can be understood from Equation 7.20.

Instead of extending the obstacle representation by the ego vehicle geometry, we can alternatively shift the vehicle origin to a point on the vehicle geometry, which is closest to an obstacle after transformation U_k. This idea is shown in Fig. 7.6. Similar to the obstacle extension approach, this ensures that c_k remains the dominating factor in Equation 7.19. Thus, $C_{obstacle} \cap B_{c_k}(\mu_k, \Sigma_k) = \emptyset$ is valid, as c_k is the maximum factor to scale $B_1(\mu_k, \Sigma_k)$.

7.2.4 Collision detection algorithm

The probability for collision avoidance as shown in Equation 7.19 is based on the factor c_k, which is derived using a distance function d. A Gilbert-Johnson-Keerthi algorithm is chosen to ensure it does not depend on a fixed set of geometric structures. Instead, GJK works with any two-dimensional convex set and determines minimum distances to the objects. It is necessary to ensure that all sets have a convex shape. Hence, each set is checked for this property, and if it does not hold, the set is partitioned into n convex subsets U_i, where the non-convex set $K = \bigcup_{i=1,...,n} U_i$.

The distance calculation is then performed for each subset to find the minimal distance of the parent sets.

As mentioned, collision checking is done using the Gilbert–Johnson–Keerthi (GJK) distance algorithm [111]. It is an efficient way to calculate the distance between two convex sets. Unlike comparable algorithms, it does not depend on the format of the geometric data. It is used as a truth value to rule out any collisions before calculating the distances. The collision check uses the Minkowski difference of two sets A and B (see Equation 7.22). If the origin lies in the resulting set of the *Minkowski difference*, this means that both sets intersect at one point at least. There is no collision when the origin is not part of the Minkowski difference.

Definition 1 Let A and B be two point sets in $\mathbb{E}_n$. The Minkowski difference $A \ominus B$ of the point sets A and B is

$$A \ominus B = \bigcap_{b \in B} A^{-b}, \tag{7.22}$$

where $A^{-b} = \{a - b | a \in A\} = A - b$ is the set A translated by $-b$

The distance is determined by a similar method. The check is twofold as shown in Equation 7.24:

Collision free We can now calculate the distance between the two objects in exactly the same way.

Collision The minimum distance of the Minkowski difference to the origin equals the minimal distance between two convex sets.

$$\text{Minimum distance between set A and B} = min\Big(d\big(p_i, (0,0)\big)\Big), p_i \in \{A \ominus B\} \quad (7.23)$$

$$\begin{aligned}(0,0) &\in A \ominus B \to A \cap B \neq \emptyset \text{ (collision)} \\ (0,0) &\notin A \ominus B \to A \cap B = \emptyset \text{ (collision free)}\end{aligned} \quad (7.24)$$

At the end of each run, the whole set of trajectories has been investigated by this method. A threshold distance determines what level of certainty is needed to keep a trajectory as final solution. The trajectory with minimal costs is then selected for the execution and motion control.

7.3 Simulation experiments

In this section we discuss whether planning with a receding horizon can be supported by treating the motion uncertainty of the ego vehicle as part of the planning system. The assumption is that this can decrease the number of re-init calls. All tests were applied in *Virtual Test Drive (VTD)*. VTD allows test loops with different levels of abstraction. The most lightweight setup for developers generates basic inputs such as occupancy grid, dynamic object list and RoadGraph from VTD's world model.

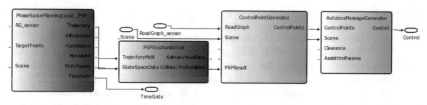

Figure 7.7: ADTF filter graph with post-processing LQG method connected to a planning module using the PSP library.

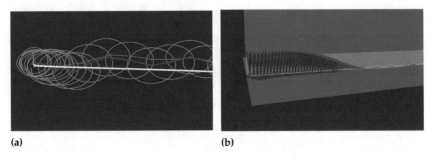

(a)   (b)

Figure 7.8: Visualization of collision probabilities: Even though predictions of lead vehicles are assumed to be correct, the ego motion in velocity space can overshoot
or undershoot an estimated state. Modeling noise adds a safety cushion to the
velocity profile.

For this experiment, control inputs are distorted by Gaussian noise $\varepsilon_n = \left(\varepsilon_\kappa^{(n)}, \varepsilon_a^{(n)} \right)$,
which leads to a distribution $\varepsilon_n \sim N_2(\mathbf{0}, M)$ with $M = \begin{pmatrix} 5 \cdot 10^{-7} & 0 \\ 0 & 1 \end{pmatrix}$. Normalization
is necessary due to the difference in magnitude of $u_\kappa^{(n)}$ and $u_a^{(n)}$. The errors added to
the inputs must be comparable to ensure stability of the filter algorithm. Otherwise,
one distorted input will dominate the other.

Assigning collision probabilities

The post-processing module is added to the ADTF filter graph as a box between
the planning module using the PSP library and the motion control modules. This
setup is shown in Fig. 7.7. The newly integrated module receives an ordered set of
100 collision-free trajectories with minimal costs from PSP as input data. The LQG
method was not implemented in CUDA due to the overhead of transferring data
between the host and the device memory for a relatively small problem.

The PSPVis visualization of collision probabilities is shown in the Figs. 7.8a and
7.8b. Probabilities are indicated through coloring and object height. Green indicates
a high probability of no collision. The color red highlights a high probability of a
collision. Ellipses are projected onto the ground to illustrate position variance. The
input trajectory was marked as valid before applying the LQG uncertainty model
(collision-free). In this case, the trajectory turns out to likely lead to a collision as
this model accounts for errors during the control phase. Although predictions of
a lead vehicle are assumed to be correct, a following or proximate maneuver can
be treated by modeling ego motion uncertainty. However, modeling noise adds a
safety cushion to the velocity profile, a similar result as a potential field around the
lead vehicle, but taking into account ego motion characteristics.

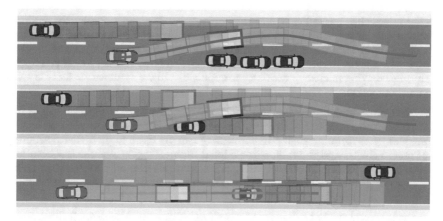

Figure 7.9: Three scenarios have been identified to benefit from modeling ego motion uncertainty: (a) Setting the right amount of lateral offset, (b) avoiding narrow passages and (c) avoiding overshooting proximate velocity profiles.

Benefits from ego motion uncertainty modeling

In summary, distorting solely a vehicle's own motion abilities leads to more conservative driving actions, as the action space is reduced and time gaps are labeled
with higher collision probabilities. It also means that fewer swerves motions are
selected as any action changing acceleration or curvature is affected by additional
noise.

One purpose was to reduce the number of re-init calls. However, this was only partially achieved in this approach. Re-inits tied to maneuvers in narrow passages did
not appear as the vehicle did not select solutions that approached these areas. On
the other hand, re-inits did continue to appear in the same frequency in interactions
with predicted dynamic obstacles. One reason could be that the prediction method
did not run its own LQG method. As soon as other vehicles showed a large difference in acceleration or relative velocities with small TTC, re-inits appeared rarely,
but in the same frequency.

In this evaluation three test cases were set up:

- Lateral offset to avoid obstacles
- Swerves in narrow passages
- Treating motion prediction errors

The illustrations are shown in Fig. 7.9a, 7.9b and 7.9c. First, Figure 7.9a illustrates
a lateral offset scenario in which the planning corridor of valid motions shrinks
significantly due to other traffic participants driving with greatly differing relative

speeds as well as partially in their neighboring lanes. Without modeling the motion uncertainty of the ego vehicle, these narrow corridors can be traversed without a collision by the optimal solution, but in further planning cycles the solution might be unrealistic for the vehicle to execute. Candidates operating in narrow passages during a driving maneuver are likely to collide due to the assumed variance in the vehicle's motion.

Second, an overtaking maneuver is evaluated in Figure 7.9b. With larger values in the curvature profile the position variance increases in the estimation by the LQG method. The former collision-free trajectory can get close to predicted positions of other vehicles and is therefore considered unsafe. In this case the LQG can be seen as an trajectory dependent safety-cushion, which acts only related to known sources of position uncertainty. In contrast to using a fix safety-area around this vehicle, less trajectories get eliminated.

In the last example the challenges of overshooting proximity behavior is illustrated in Fig. 7.9c. Although, sensor measurement error are taking into account the main remaining challenge is the existing uncertainty about the future behavior of other traffic participants. Especially for this use case it has been observed that it is not enough to cover ego motion uncertainty, but estimate an error of the prediction model as well. Ideally, this method would consider context as well. The problem of prediction is hard as velocity and acceleration profiles look similar in traffic jams and when starting at a traffic light. However, the future prediction for both cases is contradictory as the traffic jam remains a stop-and-go sequence.

As shown in earlier evaluation in Chapter 6 the planning cycle takes up to 90ms. The 10 best solution are further post-processed with the described LQG method, which takes additional 120ms on average. In this setup the planning can provide results in 4 and 5 Hz. One drawback of this method is its high dependency on available sensor information with respect to work load. The workload increases by a factor of 50 in urban environment compared to a highway scenario. This is due to detailed information on static obstacles from the occupancy grid and free space bands.

To conclude, terrain, latencies or measurement errors can lead to a discrepancy between ego motion prediction and the real vehicle position in the planning system. Modeling ego motion uncertainty help to choose trajectories that are less likely to be invalided within in the next planning cycles due to new measurements. However, the model needs to be further optimized in terms of run time and motion certainty of other traffic participants in order to add more value to the system.

8 Summary, outlook and contributions

The objective of this thesis has been to generate universal driving strategies to minimize the need for multiple special purpose planning methods. A proposal for a centralized decision-making entity has been provided as well as benchmarking criteria. The thesis provides a set of simple but adaptable driving behaviors. The implementation has been done for CPU and GPU platforms and experiments on both systems have been evaluated.

In Chapter 4, the philosophy and architecture of the *Phase Space Planning (PSP)* system is outlined. The system is universally applicable in on-road scenarios and therefore not dependent on any high-level navigation state machine for driving strategy selection. At present, partial automation products provide the customer with individual driving tasks (e.g. lane keeping or adaptive cruise control). In the future, higher degrees of automation will require additional driving modes unified in a single solution. Hence, the total number of specialized driving systems (e.g. Advanced Driver Assistant Systems) in the vehicle will decrease. Chapter 5 is dedicated to the PSP sampling-based planning approach. Implementations of state space exploration, trajectory generation and search space optimization methods have been introduced. Randomized and knowledge-based sampling strategies for state space exploration have been highlighted. The trajectory generation is based on a sophisticated method by Kelly and Nagy [62] and has been extended and optimized by universal driving strategies and GPU usage. For trajectory optimization, CPU as well as optimized GPU implementations have been presented. The adjustments to the algorithm for heavy task parallelization have been discussed. Driving modes can be combined in this planning solution, which is an advantage over independent, single-task automation systems. The most valuable source of information for an automated vehicle are its own on-board sensors. In Chapter 6, a method has been presented that improves perception quality through smart vehicle positioning. Increasing in-vehicle knowledge supports the decision-making processes and their independence from external data sources such as V2X and backend systems. These providers can then be integrated as optional information sources. Similarly, it is important to ensure stability of planning solutions and driving decisions. A linear-quadratic Gaussian (LQG) approach based on the work of van den Berg et al. [109] filters trajectories that are likely to be in collision in the event of ego motion uncertainty. Motion uncertainty is caused by terrain, system latencies and measurement errors. All lead to a planning prediction discrepancy. This stochastic optimal control mechanism is introduced in Chapter 7 and integrated into the system as a post-processing method.

The implemented modules within the PSP framework allow to generate a universal driving strategy out of four basic driving mode categories. Each one is evaluated

© Springer Fachmedien Wiesbaden GmbH, part of Springer Nature 2018
S. Heinrich, *Planning Universal On-Road Driving Strategies for Automated Vehicles*, AutoUni – Schriftenreihe 119,
https://doi.org/10.1007/978-3-658-21954-3_8

in simulation. The categories consist of start, follow and proximity behavior, double lane changes as well as basic lane change maneuvers and the scenario of taking turns. It has, therefore, been possible to show that the system complexity can be reduced to a single state planner. Hence, maintenance and parametrization are shifted from a state machine into one general objective function. These basic behaviors can be combined and executed simultaneously or as a deliberate planning sequence.

Additionally, it was shown that motion planning can create benefits for perception modules through smart positioning to increase the system's local knowledge. Two measures have been created - the area of interest entropy (AIE) and the field of view coverage of AOI (FVCAI). Minimizing AIE leads to immediate gain of knowledge in relevant areas. Therefore, the trajectory is adjusted such that it passes as many way points with small AIE values as possible without violating other components of the objective function.

Modules of PSP have been partially or fully implemented as CUDA kernels for GPUs. The minimal goal for planning in high dimensional state spaces is set to 2Hz. This is considered a minimum requirement for a real-time application in the evaluation. For the initial CPU-only implementation the requirement could not be met, when optimizing a set of 50000 trajectories or more. The run time was over 900 ms. The GPU-only implementation in comparison runs with over 10Hz and scales well in the same setup with respect to the number of samples. The workload is optimized for all implementations using profiling tools to identify hot spots, such as the calculation of the Jacobian matrix for the gradient descent used for trajectory generation.

In addition, it was shown that modeling the uncertainty of the vehicles motion supports the lifetime of planning results. A linear quadratic Gaussian (LQG) method adds a spatial safety zone around the vehicle based on the input variables of the trajectory. This is done by estimating the likelihood of a future collision given the sensor measurements and the future trajectory. Three cases have been investigated: (1) Lateral offset to avoid obstacles, (2) swerves in narrow passages and (3) treating motion prediction errors. This method adds 120ms to the overall planning cycle. For this setup, PSP can provide results in a frequency between 4 and 5 Hz. One purpose to implement the method is to reduce the number of re-init calls. The goal is partially achieved with this approach, however, motion uncertainty of traffic participants needs to be included into the model as well.

Outlook

Throughout this thesis, brief remarks have been made about future work and unanswered questions in the field. Future extensions and improvements to the current state of the art are presented in this section:

Reference path uncertainty Lane fusion algorithms using camera or lidar sensors estimate the future curvature of the road. This can be supported by localization and digital map information, although position and lane-marking extraction will have a certain degree of uncertainty. The current system operates on the latest and best lane hypothesis. For increasing covariance and more than one hypothesis, motion planning can take actions to keep several hypotheses valid until supporting measurements improve lane estimation.

Stochastic traffic prediction Future poses of other traffic participants can be predicted using the LGQ method of Chapter 7 with minor adjustments. This is partially covered by the publication of the authors Xu et al. [108] and can be greatly extended to more complex motion prediction models.

Crossing and swerving in oncoming traffic The philosophy of PSP excludes complex turning (left or U-turn) scenarios at intersections by design. Whenever knowledge has to be gained during the maneuver execution (e.g. slowly driving into an intersection to sense for oncoming traffic) this method cannot be applied. An unanswered question is how to integrate such complex but static task sequences of driving into the existing architecture of a universal planner.

Landmark-focused sensor coverage The idea of maximizing the gain in perception information while driving can be extended further. It can be used by other world modeling modules to decrease their estimation uncertainties where necessary. As one application, the localization module can actively request an urgent update on features (e.g. poles, trees etc.) to improve the position accuracy. Another aspect is an automatic identification of perception relevant areas, e.g. at intersections and turns.

Automatic cost function parametrization In PSP, all driving behaviors are shaped by an objective function as a sum of many independent cost terms. Each term has a weighting which can be optimized with respect to the overall behavior.

With self-driving cars, most extensions touch on more than one discipline. The interaction between modules such as perception, localization and motion planning becomes more important, as shown in Chapter 6. To solve these challenges at a higher level, it is necessary to integrate feedback loops across the system's architecture.

Summary of contributions

The contribution of this thesis towards motion planning for automated vehicles can be summarized as follows:

- The PSP motion planning framework uses modern GPGPUs to accelerate the task of trajectory optimization for automated vehicles. It generates universal driving strategies for the general purpose of on-road driving. The framework integrates into the existing automated driving architecture of *Volkswagen Group Research*. Planning is performed in a 7D state space (position, velocity and acceleration) along a receding temporal planning horizon.

- The philosophy of the PSP system does not rely on any higher level state automaton. Decisions are made on the basis of a set of rules and heuristics. Each planning system implemented in PSP consists of a single core state (*drive*). A major objective has been to unify planning methods and to combine their abilities. Each behavior is derived from related objective functions and cost terms.

- A smart positioning method has been developed as part of this thesis. This method supports the perception task establishing a feedback loop to the motion planning components. Similar to the philosophy of *active vision*, the vehicle is positioned to maximize sensor coverage.

- Uncertainty of ego motion can undermine the best planning intentions. Terrain, latencies or control issues can cause discrepancies between the planning input and the resulting vehicle state. This thesis has presented a post-processing method for temporal robustness. Trajectories which are potentially affected are identified using a *linear-quadratic Gaussian method (LQG)* approach to estimate the likelihood of future collisions taking into account motion uncertainty.

References

[1] *Passenger transport statistics*, Accessed: May 31, 2017. [Online]. Available: `http://ec.europa.eu/eurostat/statistics-explained/index.php/Passenger_transport_statistics#Modal_split`.

[2] World Health Organization, *Global status report on road safety 2015*. World Health Organization, 2015.

[3] European Transport Safety Counsil, *9th Annual Road Safety Performance Index Report*. European Transport Safety Counsil, 2015.

[4] SAE, *J 3016: Taxonomy and definitions for terms related to on-road motor vehicle automated driving systems*, 2014.

[5] INRIX, *Inrix 2015 traffic scorecard*, (Accessed: May 31, 2017), 2015. [Online]. Available: `inrix.com/scorecard`.

[6] T. A. T. Institute, *2015 urban mobility scorecard*, 2015.

[7] H. Moravec, *Mind children*. Cambridge Univ Press, 1988, vol. 375.

[8] S. Heinrich, R. Rojas and A. Bartels, 'Modelling complex behaviors for automated cars in phase space', in *FISITA Automotive World Congress Proceedings*, 2014.

[9] S. Heinrich, A. Zoufahl and R. Rojas, 'Real-time trajectory optimization under motion uncertainty using a gpu', in *Proceedings of the 2015 IEEE International Conference on Intelligent Robots and Systems (IROS)*, 2015.

[10] S. Heinrich, A. Bartels and R. Rojas, 'Framework zur planung einer fahrstrategie für automatische fahrzeuge in echtzeit', AAET, 2016.

[11] S. Heinrich, J. Stubbemann and R. Rojas, 'Optimizing a driving strategy by its sensor coverage of relevant environment information', in *IEEE Intelligent Vehicles Symposium*, 2016.

[12] R. Bellman, *Dynamic Programming*, 1st ed. Princeton, NJ, USA: Princeton University Press, 1957.

[13] J. P. Laumond, *Robot Motion Planning and Control*, ser. 0170-8643. Springer Berl, 1998, vol. 229.

[14] L. Kavraki, P. Svestka, J. claude Latombe and M. Overmars, 'Probabilistic roadmaps for path planning in high-dimensional configuration spaces', in *IEEE International Conference on Robotics and Automation*, 1996, pp. 566–580.

[15] N. J. Nilsson, 'A mobile automaton: An application of artificial intelligence techniques', DTIC Document, Tech. Rep., 1969.

[16] C. Ó'Dúnlaing and C. K. Yap, 'A 'retraction' method for planning the motion of a disc', *Journal of Algorithms*, vol. 6, no. 1, pp. 104–111, 1985.

© Springer Fachmedien Wiesbaden GmbH, part of Springer Nature 2018
S. Heinrich, *Planning Universal On-Road Driving Strategies for Automated Vehicles*, AutoUni – Schriftenreihe 119,
https://doi.org/10.1007/978-3-658-21954-3

[17] J. T. Schwartz and M. Sharir, 'On the 'piano movers' problem. ii. general techniques for computing topological properties of real algebraic manifolds', *Advances in applied Mathematics*, vol. 4, no. 3, pp. 298–351, 1983.

[18] R. A. Brooks and T. Lozano-Perez, 'A subdivision algorithm in configuration space for findpath with rotation', DTIC Document, Tech. Rep., 1982.

[19] J.-C. Latombe, *Robot Motion Planning*. Springer, 1991.

[20] S. F. Frisken and R. N. Perry, 'Simple and efficient traversal methods for quadtrees and octrees', *Journal of Graphics Tools*, vol. 7, no. 3, pp. 1–11, 2002.

[21] O. Khatib, 'Real-time obstacle avoidance for manipulators and mobile robots', *The international journal of robotics research*, vol. 5, no. 1, pp. 90–98, 1986.

[22] Road and Transportation Research Association (FGSV), *Guidelines for the Design of Motorways (RAA 08)*, Translated 2011, 2008.

[23] Forschungsgesellschaft für Straßen- und Verkehrswesen, *Richtlinien für die Anlage von Landstraßen (RAL 2012)*, 2012.

[24] Road and Transportation Research Association (FGSV), *Directives for the Design of Urban Roads (RASt 06)*, Translated 2011, 2006.

[25] M. Düring and P. Pascheka, 'Cooperative decentralized decision making for conflict resolution among autonomous agents', in *Innovations in Intelligent Systems and Applications (INISTA) Proceedings, 2014 IEEE International Symposium on*, IEEE, 2014, pp. 154–161.

[26] H Nagel and W Enkelmann, 'Generic road traffic situations and driver support systems', in *PROMETHEUS (PROGRAM) WORKSHOP (5TH: 1991:*, 1991.

[27] H.-H. Nagel, 'A vision of 'vision and language'comprises action: An example from road traffic', *Artificial Intelligence Review*, vol. 8, no. 2-3, pp. 189–214, 1994.

[28] W. Tölle, 'Ein fahrmanöverkonzept für einen maschinellen kopiloten', *Fortschritt Berichte - VDI Reihe 12 - Vehrkehrstechnik Fahrzeugtechnik*, 1996.

[29] S. M. LaValle, *Planning algorithms*, 2004.

[30] E. Donges, 'Aspekte der aktiven sicherheit bei der führung von personenkraftwagen', *Automobilindustrie*, vol. 27, pp. 183–190, 1982.

[31] P. Furgale, U. Schwesinger, M. Rufli, W. Derendarz, H. Grimmett, P. Muehlfellner, S. Wonneberger, J. Timpner, S. Rottmann, B. Li et al., 'Toward automated driving in cities using close-to-market sensors: An overview of the v-charge project', in *Intelligent Vehicles Symposium (IV), 2013 IEEE*, IEEE, 2013, pp. 809–816.

[32] M. Aeberhard, S. Rauch, M. Bahram, G. Tanzmeister, J. Thomas, Y. Pilat, F. Homm, W. Huber and N. Kaempchen, 'Experience, results and lessons learned from automated driving on germany's highways', *IEEE Intelligent Transportation Systems Magazine*, vol. 7, no. 1, pp. 42–57, 2015.

[33] J. Ziegler, P. Bender, M. Schreiber, H. Lategahn, T. Strauss, C. Stiller, T. Dang, U. Franke, N. Appenrodt, C. G. Keller *et al.*, 'Making bertha drive - an autonomous journey on a historic route', *IEEE Intelligent Transportation Systems Magazine*, vol. 6, no. 2, pp. 8–20, 2014.

[34] T. Victor, M. Rothoff, E. Coelingh, A. Ödblom and K. Burgdorf, 'When autonomous vehicles are introduced on a larger scale in the road transport system: The drive me project', in *Automated Driving*, Springer, 2017, pp. 541–546.

[35] M. Bojarski, D. Del Testa, D. Dworakowski, B. Firner, B. Flepp, P. Goyal, L. D. Jackel, M. Monfort, U. Muller, J. Zhang *et al.*, 'End to end learning for self-driving cars', 2016.

[36] A. Elfes, 'Using occupancy grids for mobile robot perception and navigation', *IEEE Computer*, vol. 22, no. 6, pp. 46–57, Jun. 1989.

[37] F. von Hundelshausen, M. Himmelsbach, F. Hecker, A. Müller and H.-J. Wünsche, 'Driving with tentacles: Integral structures for sensing and motion', *J. Field Robotics*, vol. 25, no. 9, pp. 640–673, 2008.

[38] H. Mouhagir, R. Talj, F. Cherfaoui Véroniqueand Guillemard and F. Aioun, 'A markov decision process-based approach for trajectory planning with clothoid tentacles', in *IEEE Intelligent Vehicles Symposium*, 2016.

[39] S. Scholz, S. Chlosta, S. Freter and F. Schuldt, 'Prototypische umsetzung eines bau- und engstellenassistenten', AAET, 2014.

[40] M. Rufli and R. Siegwart, 'On the design of deformable input- / state-lattice graphs.', in *ICRA*, IEEE, 2010, pp. 3071–3077.

[41] M. Werling, J. Ziegler, S. Kammel and S. Thrun, 'Optimal trajectory generation for dynamic street scenarios in a frenet frame', in *Robotics and Automation (ICRA), 2010 IEEE International Conference on*, IEEE, 2010, pp. 987–993.

[42] M. McNaughton, C. Urmson, J. M. Dolan and J. woo Lee, 'Motion planning for autonomous driving with a conformal spatiotemporal lattice', in *Robotics and Automation (ICRA), IEEE International Conference*, vol. 1, 2011, 4889–4895.

[43] T. Gu and J. M. Dolan, 'On-road motion planning for autonomous vehicles', in *Intelligent Robotics and Applications*, Springer, 2012, pp. 588–597.

[44] T. Gu, J. Atwood, C. Dong, J. M. Dolan and J.-W. Lee, 'Tunable and stable real-time trajectory planning for urban autonomous driving', in *Intelligent Robots and Systems (IROS), 2015 IEEE/RSJ International Conference on*, IEEE, 2015, pp. 250–256.

[45] W. Xu, J. Wei, J. M. Dolan, H. Zhao and H. Zha, 'A real-time motion planner with trajectory optimization for autonomous vehicles.', in *ICRA*, IEEE, 2012, pp. 2061–2067, ISBN: 978-1-4673-1403-9.

[46] J. Ziegler and C. Stiller, 'Spatiotemporal state lattices for fast trajectory planning in dynamic on-road driving scenarios', in *Intelligent Robots and Systems, 2009. IROS 2009. IEEE/RSJ International Conference on*, IEEE, 2009, pp. 1879–1884.

[47] C. Stiller and J. Ziegler, 'Situation assessment and trajectory planning for annieway', in *IEEE/RSJ International Conference on Intelligent Robots and Systems, Workshop on Perception and Navigation for Autonomous Vehicles in Human Environment, San Francisco, USA*, 2011.

[48] M. Pivtoraiko, R. A. Knepper and A. Kelly, 'Differentially constrained mobile robot motion planning in state lattices', *Journal of Field Robotics*, vol. 26, no. 3, pp. 308–333, 2009.

[49] M. Pivtoraiko and A. Kelly, 'Kinodynamic motion planning with state lattice motion primitives', in *2011 IEEE/RSJ International Conference on Intelligent Robots and Systems*, IEEE, 2011, pp. 2172–2179.

[50] T. M. Howard, C. J. Green, A. Kelly and D. Ferguson, 'State space sampling of feasible motions for high-performance mobile robot navigation in complex environments', *Journal of Field Robotics*, vol. 25, no. 6-7, pp. 325–345, 2008.

[51] U. Schwesinger, M. Rufli, P. Furgale and R. Siegwart, 'A sampling-based partial motion planning framework for system-compliant navigation along a reference path', in *Intelligent Vehicles Symposium (IV), 2013 IEEE*, IEEE, 2013, pp. 391–396.

[52] X. Li, Z. Sun, A. Kurt and Q. Zhu, 'A sampling-based local trajectory planner for autonomous driving along a reference path', *IEEE Intelligent Vehicles Symposium*, vol. June 2014, 2014.

[53] M. Ruf, J. R. Ziehn, B. Rosenhahn, J. Beyerer, D. Willersinn and H. Gotzig, 'Situation prediction and reaction control (sparc)', in *Tagungsband 9. Workshop Fahrerassistenzsysteme*, 2014.

[54] M. Werling, S. Kammel, J. Ziegler and L. Gröll, 'Optimal trajectories for time-critical street scenarios using discretized terminal manifolds', *I. J. Robotic Res.*, vol. 31, no. 3, pp. 346–359, 2012.

[55] B. Gutjahr and M. Werling, 'Optimale fahrzeugquerführung mittels linearer, zeitvarianter mpc', in *Tagungsband 10. Workshop Fahrerassistenzsysteme*, 2015.

[56] J. Ziegler, P. Bender, T. Dang and C. Stiller, 'Trajectory planning for bertha - a local, continuous method', in *Intelligent Vehicles Symposium Proceedings, 2014 IEEE*, IEEE, 2014, pp. 450–457.

[57] L. E. Dubins, 'On curves of minimal length with a constraint on average curvature, and with prescribed initial and terminal positions and tangents', *American Journal of mathematics*, vol. 79, no. 3, pp. 497–516, 1957.

[58] J. A. Reeds and L. A. Shepp, 'Optimal paths for a car that goes both forwards and backwards', in *Pacific Journal of Mathematics*, 1990, pp. 367–393.

[59] A. Scheuer and T. Fraichard, *Collision-free and continuous-curvature path planning for car-like robots*, 1997.

[60] T. Fraichard and A. Scheuer, 'From reeds and shepp's to continuous-curvature paths', *IEEE Transactions on Robotics*, vol. 20, no. 6, pp. 1025–1035, 2004.

[61] P. A. Theodosis and J. C. Gerdes, 'Generating a racing line for an autonomous racecar using professional driving techniques', in *ASME 2011 Dynamic Systems and Control Conference and Bath/ASME Symposium on Fluid Power and Motion Control*, American Society of Mechanical Engineers, 2011, pp. 853–860.

[62] A. Kelly and B. Nagy, 'Reactive nonholonomic trajectory generation via parametric optimal control', *I.J. Robotic Res.*, vol. 22, no. 7-8, pp. 583–602, 2003.

[63] R. L. Levien, 'From spiral to spline: Optimal techniques in interactive curve design', PhD thesis, University of California, Berkeley, 2009.

[64] M. Wang, T. Ganjineh and R. Rojas, 'Action annotated trajectory generation for autonomous maneuvers on structured road networks', in *ICARA*, IEEE, 2011, pp. 67–72, ISBN: 978-1-4577-0329-4.

[65] M. Montemerlo, J. Becker, S. Bhat, H. Dahlkamp, D. Dolgov, S. Ettinger, D. Hähnel, T. Hilden, G. Hoffmann, B. Huhnke, D. Johnston, S. Klumpp, D. Langer, A. Levandowski, J. Levinson, J. Marcil, D. Orenstein, J. Paefgen, I. Penny, A. Petrovskaya, M. Pflueger, G. Stanek, D. Stavens, A. Vogt and S. Thrun, 'Junior: The stanford entry in the urban challenge.', *J. Field Robotics*, vol. 25, no. 9, pp. 569–597, 2008, TechRep zu Junior.

[66] H. Akima, 'A new method of interpolation and smooth curve fitting based on local procedures', *Journal of the ACM (JACM)*, vol. 17, no. 4, pp. 589–602, 1970.

[67] I. E. Paromtchik and C. Laugier, 'Autonomous parallel parking of a nonholonomic vehicle', in *Intelligent Vehicles Symposium, 1996., Proceedings of the 1996 IEEE*, IEEE, 1996, pp. 13–18.

[68] D. Dolgov, S. Thrun, M. Montemerlo and J. Diebel, 'Practical search techniques in path planning for autonomous driving', *Ann Arbor*, 2008.

[69] J. Ziegler and C. Stiller, 'Fast collision checking for intelligent vehicle motion planning', in *Intelligent Vehicles Symposium (IV), 2010 IEEE*, IEEE, 2010, pp. 518–522.

[70] J. Pan and D. Manocha, 'Gpu-based parallel collision detection for fast motion planning', *The International Journal of Robotics Research*, vol. 31, no. 2, pp. 187–200, 2012.

[71] A. Lawitzky, D. Althoff, C. F. Passenberg, G. Tanzmeister, D. Wollherr and M. Buss, 'Interactive scene prediction for automotive applications', in *Intelligent Vehicles Symposium (IV), 2013 IEEE*, IEEE, 2013, pp. 1028–1033.

[72] E. W. Dijkstra, 'A note on two problems in connexion with graphs', *Numerische mathematik*, vol. 1, no. 1, pp. 269–271, 1959.

[73] A. Stentz and I. C. Mellon, 'Optimal and efficient path planning for unknown and dynamic environments', *International Journal of Robotics and Automation*, vol. 10, pp. 89–100, 1993.

[74] A. Stentz *et al.*, 'The focussed d* algorithm for real-time replanning', in *IJCAI*, vol. 95, 1995, pp. 1652–1659.

[75] D. Dolgov, S. Thrun, M. Montemerlo and J. Diebel, 'Path planning for autonomous vehicles in unknown semi-structured environments', *The International Journal of Robotics Research*, vol. 29, no. 5, pp. 485–501, 2010.

[76] H.-G. Wahl, K.-L. Bauer, F. Gauterin and M. Holzäpfel, 'A real-time capable enhanced dynamic programming approach for predictive optimal cruise control in hybrid electric vehicles', in *16th International IEEE Conference on Intelligent Transportation Systems (ITSC 2013)*, IEEE, 2013, pp. 1662–1667.

[77] M. McNaughton and C. Urmson, 'Fahr: Focused a* heuristic recomputation.', in *IROS*, Citeseer, 2009, pp. 4893–4898.

[78] S. M. LaValle and J. J. J. Kuffner, 'Randomized kinodynamic planning', *IEEE International Conference on Robotics and Automation*, 1999.

[79] X. Tang, J.-M. Lien, N. Amato *et al.*, 'An obstacle-based rapidly-exploring random tree', in *Proceedings 2006 IEEE International Conference on Robotics and Automation, 2006. ICRA 2006.*, IEEE, 2006, pp. 895–900.

[80] C. Chen, M. Rickert and A. Knoll, 'Combining space exploration and heuristic search in online motion planning for nonholonomic vehicles', in *Intelligent Vehicles Symposium (IV), 2013 IEEE*, IEEE, 2013, pp. 1307–1312.

[81] J. T. Kider, M. Henderson, M. Likhachev and A. Safonova, 'High-dimensional planning on the gpu', in *Robotics and Automation (ICRA), 2010 IEEE International Conference on*, IEEE, 2010, pp. 2515–2522.

[82] A. Bleiweiss, 'Gpu accelerated pathfinding', in *Proceedings of the 23rd ACM SIGGRAPH/EUROGRAPHICS symposium on Graphics hardware*, Eurographics Association, 2008, pp. 65–74.

[83] J. Hudecek and L. Eckstein, 'Vom reaktiven zum taktischen trajektorienplaner', in *10. Workshop Fahrerassistenzsysteme*, p. 85.

[84] T. Lee and Y. J. Kim, 'Gpu-based motion planning under uncertainties using pomdp', 2013.

[85] A. J. Davison and D. W. Murray, 'Simultaneous localization and map-building using active vision', *Pattern Analysis and Machine Intelligence, IEEE Transactions on*, vol. 24, no. 7, pp. 865–880, 2002.

[86] J. F. Seara and G. Schmidt, 'Intelligent gaze control for vision-guided humanoid walking: Methodological aspects', *Robotics and Autonomous Systems*, vol. 48, no. 4, pp. 231–248, 2004.

[87] A. Saffiotti and K. LeBlanc, 'Active perceptual anchoring of robot behavior in a dynamic environment', in *Robotics and Automation, 2000. Proceedings. ICRA'00. IEEE International Conference on*, IEEE, vol. 4, 2000, pp. 3796–3802.

[88] S. Kohlbrecher, A. Stumpf and O. von Stryk, 'Grid-based occupancy mapping and automatic gaze control for soccer playing humanoid robots', in *Proc. 6th Workshop on Humanoid Soccer Robots at the 2011 IEEE-RAS Int. Conf. on Humanoid Robots*, Bled, 2011.

[89] A. Unterholzner, M. Himmelsbach and H.-J. Wuensche, 'Active perception for autonomous vehicles', in *Robotics and Automation (ICRA), 2012 IEEE International Conference on*, IEEE, 2012, pp. 1620–1627.

[90] A. Unterholzner and H.-J. Wuensche, 'Selective attention for detection and tracking of road-networks in autonomous driving', in *Intelligent Vehicles Symposium (IV), 2013 IEEE*, IEEE, 2013, pp. 277–284.

[91] K. Patel, W. Macklem, S. Thrun and M. Montemerlo, 'Active sensing for high-speed offroad driving', in *IEEE International conference on Robotics and Automation*, Citeseer, vol. 3, 2005, p. 3162.

[92] M. Plavšić, G. Klinker and H. Bubb, 'Situation awareness assessment in critical driving situations at intersections by task and human error analysis', *Human Factors and Ergonomics in Manufacturing & Service Industries*, vol. 20, no. 3, pp. 177–191, 2010.

[93] J. Schmidhuber, 'Deep learning in neural networks: An overview', *Neural Networks*, vol. 61, pp. 85–117, 2015.

[94] J. Zhang and K. Cho, 'Query-efficient imitation learning for end-to-end autonomous driving', 2016.

[95] E. Santana and G. Hotz, 'Learning a driving simulator', 2016.

[96] R. J. Sethi and A. K. Roy-Chowdhury, 'Modeling multi-object activities in phase space', in *Asian Conference on Computer Vision*, Springer, 2010, pp. 328–337.

[97] K. Homeier and L. Wolf, 'Roadgraph: High level sensor data fusion between objects and street network', in *Intelligent Transportation Systems (ITSC), 2011 14th International IEEE Conference on*, IEEE, 2011, pp. 1380–1385.

[98] D. Topfer, J. Spehr, J. Effertz and C. Stiller, 'Efficient road scene understanding for intelligent vehicles using compositional hierarchical models', *Intelligent Transportation Systems, IEEE Transactions on*, vol. 16, no. 1, pp. 441–451, 2015.

[99] M. P. Do Carmo and M. P. Do Carmo, *Differential geometry of curves and surfaces*. Prentice-hall Englewood Cliffs, 1976, vol. 2.

[100] TC204/WG14, *Iso 15622:2010 transport information and control systems – adaptive cruise control systems – performance requirements and test procedures*, ISO, 2010.

[101] J. K. Kuchar and L. C. Yang, 'A review of conflict detection and resolution modeling methods', *IEEE Transactions on Intelligent Transportation Systems,* vol. 1, no. 4, pp. 179–189, 2000.

[102] P. Sebah and X. Gourdon, *Newton's method and high order iterations,* 2001.

[103] R. A. Knepper and A. Kelly, 'High performance state lattice planning using heuristic look-up tables', in *IROS,* 2006, pp. 3375–3380.

[104] I. Asimov, *I, robot (a collection of short stories originally published between 1940 and 1950),* 1968.

[105] R. Rojas, *I, car: The four laws of robotic cars,* (Accessed: May 31, 2017), 2011. [Online]. Available: http://inf.fu-berlin.de/inst/ag-ki/rojas_home/documents/tutorials/I-Car-Laws.pdf.

[106] D. G. Goldstein and G. Gigerenzer, 'Models of ecological rationality: The recognition heuristic.', *Psychological review,* vol. 109, no. 1, p. 75, 2002.

[107] G. Lidoris, K. Kuhnlenz, D. Wollherr and M. Buss, 'Combined trajectory planning and gaze direction control for robotic exploration', in *Robotics and Automation, 2007 IEEE International Conference on,* IEEE, 2007, pp. 4044–4049.

[108] W. Xu, J. Pan, J. Wei and J. M. Dolan, 'Motion planning under uncertainty for on-road autonomous driving', in *2014 IEEE International Conference on Robotics and Automation (ICRA),* IEEE, 2014, pp. 2507–2512.

[109] J. Van Den Berg, P. Abbeel and K. Goldberg, 'Lqg-mp: Optimized path planning for robots with motion uncertainty and imperfect state information', *The International Journal of Robotics Research,* vol. 30, no. 7, pp. 895–913, 2011.

[110] D. P. Bertsekas, *Dynamic programming and optimal control,* 2. Athena Scientific Belmont, MA, 1995, vol. 1.

[111] E. G. Gilbert, D. W. Johnson and S. S. Keerthi, 'A fast procedure for computing the distance between complex objects in three-dimensional space', *IEEE Journal on Robotics and Automation,* vol. 4, no. 2, pp. 193–203, 1988.

A Supplemental material

The appendix covers annotations to Chapters 5 and 7. In particular, the gradient descent used for path optimization and clothoid approximation (see Section 5.3.1). The results are based on related work of Kelly and Nagy [62] and McNaughton [42] and early approaches for path and velocity generation are published in Zoufahl's [1] work as part of a project advised by the author of this thesis.

Clothoid approximation for path generation

Path Jacobian

As shown in Section 5.3.1, the Jacobian of this path generation problem has the following shape:

$$J_p(\widehat{z}_p(s_g)) = \begin{pmatrix} \frac{\partial x_p}{\partial p_1} & \frac{\partial x_p}{\partial p_2} & \frac{\partial x_p}{\partial s_g} \\ \frac{\partial y_p}{\partial p_1} & \frac{\partial y_p}{\partial p_2} & \frac{\partial y_p}{\partial s_g} \\ \frac{\partial \theta_p}{\partial p_1} & \frac{\partial \theta_p}{\partial p_2} & \frac{\partial \theta_p}{\partial s_g} \end{pmatrix} \tag{A.1}$$

The Jacobian is the base for the numerical approximation of the spiral parameter set $p = [p_0, p_1, p_2, p_3, s_g]$. For each form of partial derivative an example is given in this appendix.

Spatial dimensions x_p, y_p

The equations for the vehicle position x and y coordinates include so-called Fresnel integrals. Fresnel Integrals do not have a closed form derivative and therefore cannot be solved analytically. Instead, the terms are approximated by using the Simpson rule and further differentiated. The following instructions are used to solve for x, however, the procedure to solve for y is analogue to this.

$$\begin{aligned} x_p(s) &= \int_0^s \cos(\theta_p(s))ds = \int_0^s f_p(s)ds \\ &\approx \frac{h}{3}\left(f_p(s_0) + 4f_p(s_1) + 2f_p(s_2) + \cdots + 2f_p(s_{n-2}) + 4f_p(s_{n-1}) + f_p(s_n)\right) \tag{A.2} \\ &\text{with } f_p(s) = \cos(\theta_p(s)) \quad \text{and} \quad h = \frac{s-0}{n} \end{aligned}$$

---

[1] A. Zoufahl, Real-time sampling-based trajectory generation in highway driving scenarios, 2015 (supervised by Steffen Heinrich)

© Springer Fachmedien Wiesbaden GmbH, part of Springer Nature 2018
S. Heinrich, *Planning Universal On-Road Driving Strategies for Automated Vehicles*, AutoUni – Schriftenreihe 119,
https://doi.org/10.1007/978-3-658-21954-3

Similar, $y_p(s)$ can be derived:

$$
\begin{aligned}
y_p(s) &= \int_0^s \sin(\theta_p(s))ds = \int_0^s g_p(s)ds \\
&\approx \frac{h}{3}\left(g_p(s_0) + 4g_p(s_1) + 2g_p(s_2) + \cdots + 2g_p(s_{n-2}) + 4g_p(s_{n-1}) + g_p(s_n)\right) \quad \text{(A.3)} \\
&\text{with } g_p(s) = \sin(\theta_p(s)) \quad \text{and} \quad h = \frac{s-0}{n}
\end{aligned}
$$

The spiral is defined as curvature κ over station s, therefore, the position can be derived by using the orientation θ in the Fresnel integral.

Partial derivatives for spatial dimensions

Given $x_p(s)$ and $y_p(s)$, the partial derivatives for the Jacobian can be calculated.

$$
\begin{aligned}
\frac{\partial}{\partial p_1} x_p(s_g) &= \frac{h}{3} \cdot \frac{\partial}{\partial p_1}\left(f_p(s_0) + 4f_p(s_1) + \cdots + 4f_p(s_{n-1}) + f_p(s_n)\right) \\
&= \frac{h}{3}\left(\frac{\partial}{\partial p_1} f_p(s_0) + 4\frac{\partial}{\partial p_1} f_p(s_1) + \cdots\right)
\end{aligned}
\quad \text{(A.4)}
$$

In the following equation f_p is replaced by $g_p = \sin(\theta_p(s))$ through differentiation.

$$
\frac{\partial}{\partial p_1} f_p(s_k) = \frac{\partial}{\partial p_1}\cos(\theta_p(s_k)) \quad \text{(A.5)}
$$

$$
= -\sin(\theta_p(s_k)) \cdot \frac{\partial}{\partial p_1}\theta_p(s_k) \quad \text{(A.6)}
$$

$$
= -g_p(s_k) \cdot \theta'_{p1}(s_k) \qquad \text{with } \theta'_{pj}(s_k) = \frac{\partial}{\partial p_j}\theta_p(s_k) \quad \text{(A.7)}
$$

The equation for the sum are as follows:

$$
\begin{aligned}
\frac{\partial}{\partial p_1} x_p(s_g) &= \frac{h}{3} \cdot \frac{\partial}{\partial p_1}\left(f_p(s_0) + 4f_p(s_1) + \cdots + 4f_p(s_{n-1}) + f_p(s_n)\right) \\
&= \frac{h}{3} \cdot \left(-g_p(s_0)\theta'_{p1}(s_0) - 4g_p(s_1)\theta'_{p1}(s_1) - 2g_p(s_2)\theta'_{p1}(s_2) - \right. \\
&\qquad \left. \cdots - 4g_p(s_{n-1})\theta'_{p1}(s_{n-1}) - g_p(s_n)\theta'_{p1}(s_n)\right)
\end{aligned}
\quad \text{(A.8)}
$$

In the same manner the equations for $\frac{\partial}{\partial p_2} x_p(s)$ are computed:

$$\frac{\partial}{\partial p_2} x_p(s_g) = \frac{h}{3} \cdot \frac{\partial}{\partial p_2} \left(f_p(s_0) + 4f_p(s_1) + \cdots + 4f_p(s_{n-1}) + f_p(s_n) \right)$$

$$= \frac{h}{3} \cdot \left(-g_p(s_0) \cdot \theta'_{p2}(s_0) - 4g_p(s_1)\theta'_{p2}(s_1) - 2g_p(s_2)\theta'_{p2}(s_2) - \cdots - 4g_p(s_{n-1})\theta'_{p2}(s_{n-1}) - g_p(s_n)\theta'_{p2}(s_n) \right) \quad \text{(A.9)}$$

Orientation θ_p

In contrast to the solution for spatial dimensions x and y the orientation $\theta(s)$ is the antiderivative of the curvature function $\kappa(s)$. The orientation's equation is rearranged such that the terms are ordered by coefficients of p_i, rather than the previous introduced notation as coefficients of s^i.

$$\theta_p(s) = a(p) \cdot s + b(p) \cdot \frac{s^2}{2} + c(p) \cdot \frac{s^3}{3} + d(p) \cdot \frac{s^4}{4}$$

$$= p_0 \cdot s - \frac{11p_0 - 18p_1 + 9p_2 - 2p_3}{2 \cdot s_g} \cdot \frac{s^2}{2}$$

$$+ 9 \cdot \frac{2p_0 - 5p_1 + 4p_2 - p_3}{2 \cdot s_g^2} \cdot \frac{s^3}{3} - 9 \cdot \frac{p_0 - 3p_1 + 3p_2 - p_3}{2 \cdot s_g^3} \cdot \frac{s^4}{4} \quad \text{(A.10)}$$

$$= p_0 \cdot \left(s + \frac{-11}{4} \frac{s^2}{s_g} + 3\frac{s^3}{s_g^2} + \frac{-9}{8} \frac{s^4}{s_g^3} \right) + p_1 \cdot \left(\frac{9}{2} \frac{s^2}{s_g} + \frac{-15}{2} \frac{s^3}{s_g^2} + \frac{27}{8} \frac{s^4}{s_g^3} \right)$$

$$+ p_2 \cdot \left(\frac{-9}{4} \frac{s^2}{s_g} + 6\frac{s^3}{s_g^2} + \frac{-27}{8} \frac{s^4}{s_g^3} \right) + p_3 \cdot \left(\frac{1}{2} \frac{s^2}{s_g} + \frac{-3}{2} \frac{s^3}{s_g^2} + \frac{9}{8} \frac{s^4}{s_g^3} \right)$$

Do note, that s is different from s_g in general, but is equal, if we compute the Jacobian. If $s = s_g$, the equation can be reduced to the following:

$$\theta_p(s_g) = p_0 \cdot \left(s_g + \frac{-11}{4} \frac{s_g^2}{s_g} + 3\frac{s_g^3}{s_g^2} + \frac{-9}{8} \frac{s_g^4}{s_g^3} \right) + p_1 \cdot \left(\frac{9}{2} \frac{s_g^2}{s_g} + \frac{-15}{2} \frac{s_g^3}{s_g^2} + \frac{27}{8} \frac{s_g^4}{s_g^3} \right)$$

$$+ p_2 \cdot \left(\frac{-9}{4} \frac{s_g^2}{s_g} + 6\frac{s_g^3}{s_g^2} + \frac{-27}{8} \frac{s_g^4}{s_g^3} \right) + p_3 \cdot \left(\frac{1}{2} \frac{s_g^2}{s_g} + \frac{-3}{2} \frac{s_g^3}{s_g^2} + \frac{9}{8} \frac{s_g^4}{s_g^3} \right) \quad \text{(A.11)}$$

$$= \left(p_0 \frac{1}{8} + p_1 \frac{3}{8} + p_2 \frac{3}{8} + p_3 \frac{1}{8} \right) \cdot s_g$$

The partial derivatives are now easily obtainable:

$$
\frac{\partial}{\partial p_1} \theta_p(s_g) = \frac{\partial}{\partial p_1} \left[\left(p_0 \frac{1}{8} + p_1 \frac{3}{8} + p_2 \frac{3}{8} + p_3 \frac{1}{8} \right) \cdot s_g \right]
$$
$$
= \frac{3}{8} \cdot s_g
$$
(A.12)

$$
\frac{\partial}{\partial p_2} \theta_p(s_g) = \frac{\partial}{\partial p_2} \left[\left(p_0 \frac{1}{8} + p_1 \frac{3}{8} + p_2 \frac{3}{8} + p_3 \frac{1}{8} \right) \cdot s_g \right]
$$
$$
= \frac{3}{8} \cdot s_g
$$
(A.13)

$$
\frac{\partial}{\partial s_g} \theta_p(s_g) = \frac{\partial}{\partial s_g} \left[\left(p_0 \frac{1}{8} + p_1 \frac{3}{8} + p_2 \frac{3}{8} + p_3 \frac{1}{8} \right) \cdot s_g \right]
$$
$$
= p_0 \frac{1}{8} + p_1 \frac{3}{8} + p_2 \frac{3}{8} + p_3 \frac{1}{8}
$$
(A.14)

Given the equations A.8, A.9 as well as A.12-A.14, the Jacobian can be calculated for the path generation problem.

Linear Quadratic Gaussian: Kalman filter

In Chapter 7 the process of the stochastic optimal control method is described using a Linear-Quadratic Gaussian approach. The process is has a Kalman filter at its core. The 2 stages of a Kalman filter are described in the chapter. In addition, more background information is given in the following section.

Prediction and update stages

As shown in the LQG reference implementation of van den Berg et al. [109] the phase of prediction, the corresponding covariance matrix to the estimator $\tilde{x}_{k|k-1}$ is defined as follows:

$$
P_{k|k-1} := \mathbb{E}((\tilde{x}_k - \tilde{x}_{k|k-1})(\tilde{x}_k - \tilde{x}_{k|k-1})^T).
$$

Thus, the covariance of the a priori approximation error can be written as:

$$
P_{k|k-1} = A_k P_{k-1|k-1} A_k^T + V_k M_k V_k^T,
$$
(A.15)

where $P_{k-1|k-1}$ is the covariance of an a posteriori approximation error by the estimator with respect to $\bar{x}_{k-1}$, using measurements including z_{k-1}:

$$P_{k-1|k-1} = \mathbb{E}((\bar{x}_{k-1} - \tilde{x}_{k-1|k-1})(\bar{x}_{k-1} - \tilde{x}_{k-1|k-1})^T)$$

In the phase of correction new measurements $\bar{z}_k$ are used to update the estimator. Therefore, a Kalman filter has a gain matrix. In addition to what is discussed in Chapter 7, the gain matrix modifies the covariance matrix of the last prediction step as well as the estimator:

$$K_k = P_{k|k-1} H_k^T \left(H_k P_{k|k-1} H_k^T + W_k N_k W_k^T\right)^{-1}. \tag{A.16}$$

The matrix N_k is the covariance matrix of ξ_k. The value of a measurement and prediction compound (defined by K_k) is added to the previous estimator form $\tilde{x}_{k|k}$. The covariance matrix is formed in a similar way:

$$\tilde{x}_{k|k} = \tilde{x}_{k|k-1} + K_k \left(\bar{z}_k - H_k \tilde{x}_{k|k-1}\right). \tag{A.17}$$

$$P_{k|k} = (I - K_k H_k) P_{k|k-1}. \tag{A.18}$$

A priori distribution of inputs

The derivation of an a priori distribution of inputs is shown in [109]. Using $\Lambda_k := \begin{bmatrix} I & 0 \\ 0 & L_{k+1} \end{bmatrix}$, gives a shared a priori distribution of

$$\begin{bmatrix} x_k \\ u_k \end{bmatrix} \sim N(0, \Lambda_k R_k \Lambda_k^T). \tag{A.19}$$

The entries of the sub-matrix of the covariance matrix that hold information about the x and y coordinates of state x can be used to derive the likelihood of collisions. The correctness of the result of Equation A.19 and its covariances $cov(\begin{bmatrix} x_i \\ u_i \end{bmatrix}, \begin{bmatrix} x_j \\ u_j \end{bmatrix})$ does not depend on the feasibility and observability of the dynamics, nor on the Equations 7.11 and 7.12 in Chapter 7 for the observation model.

Printed in the United States
By Bookmasters